COMPUTER AIDED CIVIL ENGINEERING DRAWING

AUTHOR (S)

Dr. D. Chandra Mouli Darapaneni
M.Tech., Ph.D.
Assistant Professor,
Department of Civil Engineering,
University College of Engineering and Technology,
Acharya Nagarjuna University

Mrs B. Swathi
M.Tech., (Ph.D.)
Assistant Professor,
Department of Civil Engineering,
University College of Engineering and Technology,
Acharya Nagarjuna University

REVIEWER

DR. D. Ravi Prasad M.Tech., Ph.D.
Associate Professor
Structures Division
Department of Civil Engineering,
NIT Warangal

KHANNA BOOK PUBLISHING CO. (P) Ltd.
4C/4344, Ansari Road, Darya Ganj,
New Delhi, India

All India Council for Technical Education
Nelson Mandela Marg, Vasant Kunj,
New Delhi, India

Dear Readers,

To prevent the piracy, this book is secured with HIGH SECURITY HOLOGRAM on the front title cover. In case you don't find the hologram on the front cover title, please write us to at contact@khannabooks.com or whatsapp us at +91-99109 09320 and avail special gift voucher for yourself.

Specimen of Hologram on front Cover title:

Moreover, there is a SPECIAL DISCOUNT COUPON for you with EVERY HOLOGRAM.

How to avail this SPECIAL DISCOUNT:

Step 1: Scratch the hologram

Step 2: Under the scratch area, your "coupon code" is available

Step 3: Logon to www.khannabooks.com

Step 4: Use your "coupon code" in the shopping cart and get your copy at a special discount

Step 5: Enjoy your reading!

ISBN: 978-93-5538-218-4

Book Code: UG418EN

Computer Aided Civil Engineering Drawing

by Dr. D. Chandra Mouli Darapaneni,
Mrs B. Swathi

First Edition: 2026

Originally Published by:
All India Council for Technical Education, Ministry of Education, GOI

Exclusively Published & Distributed by:
Khanna Book Publishing Co. (P) Ltd.
4C/4344, Ansari Road, Daryaganj,
New Delhi (INDIA)

Visit us at: **www.khannabooks.com**
Write us at: **contact@khannabooks.com**

To view complete list of books, Please scan the QR Code: KPH

Printed in Bharat.

प्रो. टी. जी. सीताराम
अध्यक्ष
Prof. T. G. Sitharam
Chairman

सत्यमेव जयते

आज़ादी का अमृत महोत्सव

अखिल भारतीय तकनीकी शिक्षा परिषद्
(भारत सरकार का एक सांविधिक निकाय)
(शिक्षा मंत्रालय, भारत सरकार)
नेल्सन मंडेला मार्ग, वसंत कुंज, नई दिल्ली–110070
दूरभाष : 011–26131498
ई–मेल : chairman@aicte-india.org

ALL INDIA COUNCIL FOR TECHNICAL EDUCATION
(A STATUTORY BODY OF THE GOVT. OF INDIA)
(Ministry of Education, Govt. of India)
Nelson Mandela Marg, Vasant Kunj, New Delhi-110070
Phone : 011-26131498
E-mail : chairman@aicte-india.org

FOREWORD

Engineers are the backbone of any modern society. They are the ones responsible for the marvels as well as the improved quality of life across the world. Engineers have driven humanity towards greater heights in a more evolved and unprecedented manner.

The All India Council for Technical Education (AICTE), have spared no efforts towards the strengthening of the technical education in the country. AICTE is always committed towards promoting quality Technical Education to make India a modern developed nation emphasizing on the overall welfare of mankind.

An array of initiatives has been taken by AICTE in last decade which have been accelerated now by the National Education Policy (NEP) 2020. The implementation of NEP under the visionary leadership of Hon'ble Prime Minister of India envisages the provision for education in regional languages to all, thereby ensuring that every graduate becomes competent enough and is in a position to contribute towards the national growth and development through innovation & entrepreneurship.

One of the spheres where AICTE had been relentlessly working since past couple of years is providing high quality original technical contents at Under Graduate & Diploma level prepared and translated by eminent educators in various Indian languages to its aspirants. For students pursuing 2nd year of their Engineering education, AICTE has identified 87 books, which shall be translated into 12 Indian languages - Hindi, Tamil, Gujarati, Odia, Bengali, Kannada, Urdu, Punjabi, Telugu, Marathi, Assamese & Malayalam. In addition to the English medium, books in different Indian Languages are going to support the students to understand the concepts in their respective mother tongue.

On behalf of AICTE, I express sincere gratitude to all distinguished authors, reviewers and translators from the renowned institutions of high repute for their admirable contribution in a record span of time.

AICTE is confident that these outcomes based original contents shall help aspirants to master the subject with comprehension and greater ease.

(Prof. T. G. Sitharam)

ACKNOWLEDGEMENT

The authors are grateful to the authorities of AICTE, particularly Prof. Prof. T. G. Sitharam; Chairman, Dr. Abhay Jere; Vice-Chairman, Prof. Rajive Kumar; Member Secretary, and Dr. Sunil Luthra; Director of Training and Learning Bureau for their planning to publish the books on ***Computer Aided Civil Engineering Drawing***. We sincerely acknowledge the valuable contributions of the reviewer of the book Dr. D. Ravi Prasad, Associate Professor, Structures Division, Department of Civil Engineering, NIT Warangal.

We hereby extend our heartfelt gratitude and appreciation to S. Naveen, G. Ganesh, S. Ramakrishna, S. Rahul, G. Jai Krishna and Sh. Jabbar for their invaluable contributions to the creation of this book. This book is an outcome of various suggestions of AICTE members, experts and authors who shared their opinion and thought to further develop the engineering education in our country. Acknowledgements are due to the contributors and different workers in this field whose published books, review articles, papers, photographs, footnotes, references and other valuable information enriched us at the time of writing the book.

Dr. D CHANDRA MOULI

Mrs. B. SWATHI

PREFACE

In the world of computers, making things easier and more efficient, drawing using a computer is the basic (fundamental) knowledge required for a student to show the visualization on a paper. The art of expressing the engineer/ architect's plan, elevation or section to the client by fulfilling the requirement of building the dream house needs a platform to communicate, which led to the development of CAD (Computer Aided Drawing).

The text book describes thoroughly the basics of drawing entities used for developing building plans, elevations, sections, and perspective views. The fundamentals required for developing structural drawings along with architectural drawings are presented keeping view on both students of architecture and civil engineering attending the college of polytechnic and engineering. Number of drawings are attached at the end for practising sessions.

Prerequisites for understanding the book are the fundamental concepts on mathematics, building materials, and knowing various elements of the building will help to understand the book in a better manner.

I am very thankful to all my Guru's who made me stand in this position and also to all my students who helped me all the way to wipe out errors and ease of understanding the book in a better way. I thank my mother, Padmavathi Darapaneni, who is the real warrior I have seen in my life, she crossed all the barriers to getting the best education and everything I have today in my life.

Dr. D CHANDRA MOULI

Mrs. B. SWATHI

OUTCOME BASED EDUCATION

For the implementation of an outcome-based education the first requirement is to develop an outcome-based curriculum and incorporate an outcome-based assessment in the education system. By going through outcome-based assessments evaluators will be able to evaluate whether the students have achieved the outlined standard, specific and measurable outcomes. With the proper incorporation of outcome-based education there will be a definite commitment to achieve a minimum standard for all learners without giving up at any level. At the end of the programme running with the aid of outcome-based education, a student will be able to arrive at the following outcomes:

PO1. Engineering knowledge: Apply the knowledge of mathematics, science, engineering fundamentals, and an engineering specialization to the solution of complex engineering problems.

PO2. Problem analysis: Identify, formulate, review research literature, and analyse complex engineering problems reaching substantiated conclusions using first principles of mathematics, natural sciences, and engineering sciences.

PO3. Design / development of solutions: Design solutions for complex engineering problems and design system components or processes that meet the specified needs with appropriate consideration for the public health and safety, and the cultural, societal, and environmental considerations.

PO4. Conduct investigations of complex problems: Use research-based knowledge and research methods including design of experiments, analysis and interpretation of data, and synthesis of the information to provide valid conclusions.

PO5. Modern tool usage: Create, select, and apply appropriate techniques, resources, and modern engineering and IT tools including prediction and modelling to complex engineering activities with an understanding of the limitations.

PO6. The engineer and society: Apply reasoning informed by the contextual knowledge to assess societal, health, safety, legal and cultural issues and the consequent responsibilities relevant to the professional engineering practice.

PO7. Environment and sustainability: Understand the impact of the professional engineering solutions in societal and environmental contexts, and demonstrate the knowledge of, and need for sustainable development.

PO8. Ethics: Apply ethical principles and commit to professional ethics and responsibilities and norms of the engineering practice.

PO9. Individual and team work: Function effectively as an individual, and as a member or leader in diverse teams, and in multidisciplinary settings.

PO10. Communication: Communicate effectively on complex engineering activities with the engineering community and with society at large, such as, being able to comprehend and write effective reports and design documentation, make effective presentations, and give and receive clear instructions.

PO11. Project management and finance: Demonstrate knowledge and understanding of the engineering and management principles and apply these to one's own work, as a member and leader in a team, to manage projects and in multidisciplinary environments.

PO12. Life-long learning: Recognize the need for, and have the preparation and ability to engage in independent and life-long learning in the broadest context of technological change.

COURSE OUTCOMES

After completion of the course the students will be able to:

CO-1: Develop parametric design and the conventions of formal engineering drawing.

CO-2: Produce and interpret 2D & 3D drawings.

CO-3: Communicate a design idea/concept graphically /visually.

CO-4: Examine a design critically and with understanding of CAD – The students learns to interpret drawings, and to produce designs using a combination of 2D & 3D Software.

CO-5: Get a detailed study of an engineering artefact.

Course Outcomes	Expected Mapping with Programme Outcomes *(1- Weak Correlation; 2- Medium correlation; 3- Strong Correlation)*											
	PO-1	**PO-2**	**PO-3**	**PO-4**	**PO-5**	**PO-6**	**PO-7**	**PO-8**	**PO-9**	**PO-10**	**PO-11**	**PO-12**
CO-1	3	-	-	-	3	-	-	2	3	2	1	2
CO-2	2	2	3	2	2	-	-	2	2	1	-	2
CO-3	2	2	2	3	3	-	-	2	3	2	1	2
CO-4	2	2	3	1	3	-	-	2	1	2	1	2
CO-5	2	2	2	1	3	-	-	2	1	2	1	2

GUIDELINES FOR TEACHERS

To implement Outcome Based Education (OBE) knowledge level and skill set of the students should be enhanced. Teachers should take a major responsibility for the proper implementation of OBE. Some of the responsibilities (not limited to) for the teachers in OBE system may be as follows:

- Within reasonable constraint, they should maneuver time to the best advantage of all students.
- They should assess the students only upon certain defined criterion without considering any other potential ineligibility to discriminate them.
- They should try to grow the learning abilities of the students to a certain level before they leave the institute.
- They should try to ensure that all the students are equipped with the quality knowledge as well as competence after they finish their education.
- They should always encourage the students to develop their ultimate performance capabilities.
- They should facilitate and encourage group work and team work to consolidate newer approach.
- They should follow Blooms taxonomy in every part of the assessment.

Bloom's Taxonomy

Level	Teacher should Check	Student should be able to	Possible Mode of Assessment
Create	Students' ability to create	Design or create	Mini project
Evaluate	Students' ability to justify	Argue or defend	Assignment
Analyze	Students' ability to distinguish	Differentiate or distinguish	Project/Lab Methodology
Apply	Students' ability to use information	Operate or demonstrate	Technical Presentation/ Demonstration
Understand	Students' ability to explain the ideas	Explain or classify	Presentation/Seminar
Remember	Students' ability to recall (or remember)	Define or Recall	Quiz

GUIDELINES FOR STUDENTS

Students should take equal responsibility for implementing the OBE. Some of the responsibilities (not limited to) for the students in OBE system are as follows:

- Students should be well aware of each UO before the start of a unit in each and every course.
- Students should be well aware of each CO before the start of the course.
- Students should be well aware of each PO before the start of the programme.
- Students should think critically and reasonably with proper reflection and action.
- Learning of the students should be connected and integrated with practical and real-life consequences.
- Students should be well aware of their competency at every level of OB

ABBREVIATIONS AND SYMBOLS

List of Abbreviations

General Terms			
Abbreviations	**Full form**	**Abbreviations**	**Full form**
Ctrl	Control	Tan	Tangent
Shft	Shift	Y	Yes
F	Function	N	No
DPC	Damp Proof Coarse	GL	Ground Level
BIM	Building Information Modelling	LEP	Left Eye Point
LVP	Left Vanishing Point	RVP	Right Vanishing point
REP	Right Eye Point	SP	Station Point
PP	Picture Plane	PL	Plinth Level

List of Symbols

Symbols	Description	Symbols	Description
Bm	Beams	Alt	Alternate bar
Col	Columns	Bt	Bent bar
Fg	Footing	B	Bottom bar
JT	Joints	St	Straight bar
LL	Lintels	min	Minimum
GR	Girders	max	Maximum
LB	Lintel Beams	Stp	Stirrup
Sb or S	Slab	Ct	Column Tie
c	Centre Line	T	Top bar
@	Spacing center to center	Sp	Spiral
Φ	Diameter of bar	#	High yield Strength bars

LIST OF FIGURES

Unit 1 Introduction

Unit 2 Symbols and Sign Conventions

Unit 5 Pictorial View

LIST OF TABLES

CONTENTS

1 INTRODUCTION

UNIT SPECIFICS

The following aspects are discussed in this module

- What is drawing, Types of drawing
- Layouts of drawing
- Introduction to AutoCAD
- Drawing tools available in CAD
- Drawing objects using CAD

RATIONALE

This chapter enables students to understand the concept and importance of drawing to communicate effectively while doing the work. A comprehensive understanding of drawing various objects with perfect scales, tools, and drawing standards are discussed.

PRE-REQUISITES

- Basic Mathematics (Relative distances, angles)
- Engineering Graphics
- Elements of Building

UNIT OUTCOMES

The list of outcomes of this unit is as follows:

U1-O1: Describe various types of drawings

U1-O2: Capable of using modify commands, drawing entities, layers, texting and standards

U1-O3: Apply various commands to develop different objects/ structures.

Unit- 1 Outcomes	EXPECTED MAPPING WITH COURSE OUTCOMES (1-Weak correlation; 2-Medium correlation; 3-Strong correlation)				
	CO-1	CO-2	CO-3	CO-4	CO-5
U1-O1	-	-	3	-	2
U1-O2	2	3	3	3	2
U1-O3	3	3	3	3	3

1.1 INTRODUCTION TO THE CONCEPT OF DRAWINGS

Drawing is the process of expressing an idea into a visual format. Drawing is simply a type of representation that engineers use to discuss their ideas about planning, designing an object, or building a structure for a particular purpose. Drawings do not require language to function.

An engineer's drawing effectively conveys an object's details without the need for a lengthy complicated description that could require a lot of words. Drawings not only depict an object's details but also its various components, position, dimension, and materials that are actually required to use it effectively.

When creating plans, sections and elevations of a structure, an architect relies on drawings to effectively depict the dimensions and positions of each room. Drawings are an efficient way for a civil engineer to illustrate all the structural parts that are required to be built.

The students of civil engineering should be capable of understanding any type of drawing with ease to effectively convert the idea of an architect and structural engineer into reality with good stability and serviceability.

1.2 INTERPRETATION OF TYPICAL DRAWINGS

One of the most widely used forms of graphic communication is the drawing. There are two basic types of drawings

(i) Artistic Drawing

(ii) Technical Drawing

Artistic Drawing

Artistic drawings range in scope from the simplest line drawing to the most famous paintings. Regardless of their complexity, artistic drawings are used to express the feeling, beliefs, philosophies and ideas of the artist.

Technical Drawing

A technical drawing is a clearly and concisely communicates all of the information necessary to transform an idea or a concept in to reality. This drawing often contains more than just a graphic representation of its subject. It also contains dimension, text and specifications. These drawings typically use standardized symbols, lines and annotations to communicate specific objects.

1.3 PLANNING DRAWINGS TO SHOW INFORMATION CONCISELY

Clarity and simplicity: Try to keep the drawings as straightforward as Possible while capturing essential details. Labels, symbols and annotations should be simple to understand

Standardization: Symbols, measurements and notations should follow industry norms and customs. This ensures that anyone who are not unfamiliar with engineering drawings may understand designs with ease

Consistency: keep the drawing set consistent from beginning to end. All drawings related to a specific project should use the same conventions symbols and units.

Use of scale: Include a scale on the drawing to indicate dimensions accurately. This helps in understanding the proportions and sizes of different components and is used to transfer an object's true or relative dimension to the drawing.

Detail and annotation: Don't overload the drawing with too many details. Additional information can be provided as needed without using much space in the main display by using annotations and notes.

Title blocks: It contains important information about the drawing providing details necessary for identification, reference and documentation purposes.

Visulization tools: To properly represent the object or system use a variety of views. Sections, details and assembly drawings can also help provide more depth and clarity.

Variety of views:

***Orthographic View:** This view in engineering drawing is used to represent 3D object in a 2D Space. They show multiple 2D views of an object from different angles and help to understand its shape size and features.

***Isometric View:** This view is used to represent a 3D object in 2D space. It is a type of graphic Representation where an object is shown on a 2D plane but forms a view Point.

1.4 OPTIMUM LAYOUT OF DRAWINGS AND SCALES

Title blocks: On every design sheet place the title block in the same spot usually in the lower right corner. It includes important details such as student, date, scale, scale of project and title.

Margins: When printing or recreating the drawing, ensure it has enough margins to prevent any information from being cut off.

Consistency: To ensure accurate understanding and measurements, utilize the same scales consistently across different viewpoints and sheets.

Scale: Select the right scale for each view or section of the drawing. Depending on the size of the object being drawn, common scales are 1:1 (true size), 1:2, 1:5, 1:10, etc.…

Drawing sheet size: Select the right drawing sheet size (A0, A1, A2, etc.…) according to the precision and size of the drawing.

Space utilization: Make the most of the available space on the design page by minimizing blank areas and ensuring readability and clarity.

Cross referencing: Make sure there is clear cross- referencing between any views or parts in your drawing to show connections and interconnections.

Annotations and notes: Add any relevant comments, symbols and notes to help clarify any particulars or instructions.

1.5 INTRODUCTION TO COMPUTER AIDED DRAWING

AutoCAD is a Computer Aided Drawing (CAD) software developed and marketed by Autodesk. AutoCAD enables Architects, designers, Engineers and Drafters to make exact and detailed drawings of things like machines, buildings etc. It provides a variety of tools for modeling, annotating and drafting, assisting experts in bringing their design perceptions of life. AutoCAD has the ability to design in both 2D and 3D. The creation of floor plans, elevations, sections and other technical drawings is possible in the 2D format. The 3D mode enables to create 3D models of objects and locations, providing a more accurate representation of designs. The graphical user interface of AutoCAD is simple to use and contains a number of menus, toolbars, and a command line interface for entering commands. The interface enables users to easily build and manipulate objects and has access to a variety of drawing and editing tools. There are several drawing tools available in AutoCAD namely lines, rectangles, circles, polygon, arc, ellipse and more. These objects can also be edited and modified using commands like rotate, trim, chamfer, fillet, mirror etc. To make it simple to communicate design requirements and measurements, AutoCAD users can add dimensions, text labels, and annotations to their drawings. Users can arrange their drawings by adding various items to layers, which can be switched on or off to regulate visibility. Each item may have unique characteristics, such as line color, line type, and line thickness. AutoCAD is compatible with several file formats, including its own. The industry generally uses the DWG format. To facilitate collaboration with users of various AutoCAD software, it can also import and export other file types.

USER INTERFACE

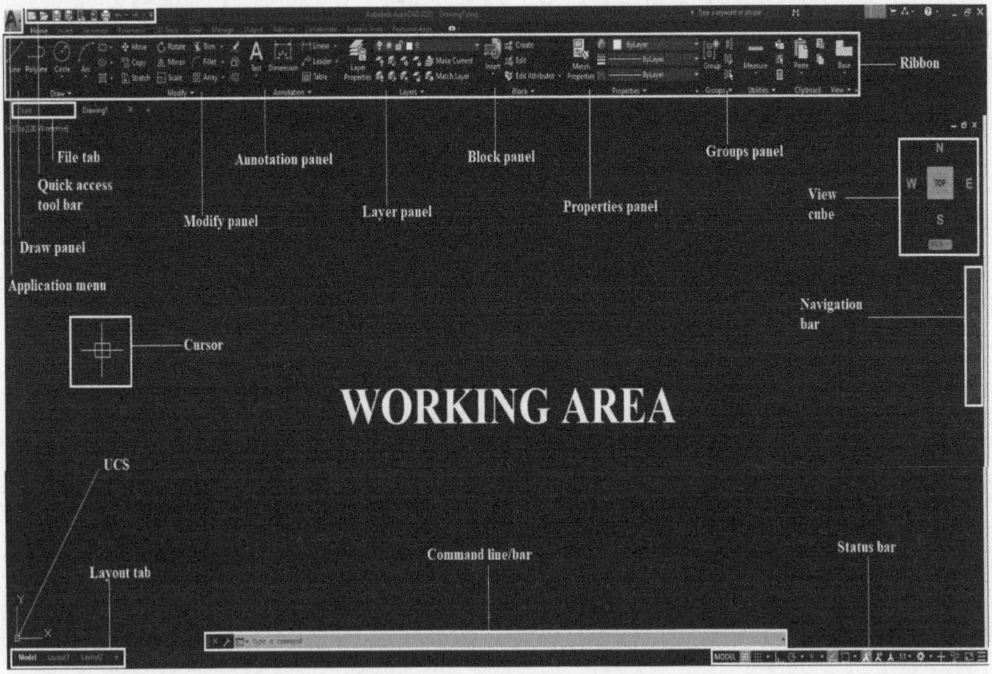

Fig. 1.1: User Interface - Components of AutoCAD

1.5.1 To Use Auto CAD Commands:

Auto CAD commands can be chosen in different ways, with the three techniques listed below being the most commonly used.

(a) Type the command through the keyboard
(b) Select the command icon from the toolbar
(c) Select the command from the screen menu

1.5.2 APPLICATION MENU:

The application menu is an important aspect of the user interface as it provides access to a wide range of commands and features. The following are the useful commands in the AutoCAD application menu

COMMAND: NEW (or) **Ctrl +N**

The NEW command works to generate a new, blank drawing file. This command allows the user to define drawing parameters such as a drawing template, units of measurement, and other drawing parameters.

COMMAND: OPEN (or) **Ctrl +O**

To open a previously saved AutoCAD drawing file, use this command. When choosing a file to open you are promoted to travel to that location on your computer or network.

COMMAND: CLOSE (or) **Ctrl +C**

This command is used to close the presently active drawing file without leaving AutoCAD.

COMMAND: SAVE (or) **Ctrl +S**

A Drawing file that is open at the moment can be saved with the Save command.

COMMAND: SAVEAS (or) **Ctrl+Shft+S**

In AutoCAD the Save as command is used to save the present drawing file under an alternative name or in a different place. This command is very helpful for duplicating an existing drawing while maintaining the original.

COMMAND: PRINT (or) **Ctrl+P**

The print command opens the plot dialog box. Users can use this Ctrl+P shortcut to rapidly access the plot settings and options for printing or plotting drawings in AUTOCAD.

COMMAND: ZOOM (or) **Z**

Zoom command is used to enlarge or reduce the display of a drawing.

COMMAND: PAN (or) **P**

Pan allows you to change the view of your work without changing the magnification factor.

COMMAND: UNDO (or) **Ctrl +Z**

This command is used to restore the previous action. For example, if an object is deleted in the drawing accidentally then one can restore the object by using UNDO command.

COMMAND: REDO (or) **Ctrl +Y**

Reverses the previous UNDO or U command. For example, to delete an object that was restored by using UNDO command, it can be deleted by using REDO command.

1.5.3 KEYBOARD SHORTCUTS

Ctrl+0	To clear the screen
Ctrl+1	To open the properties palette
Ctrl+2	To design the centre palette
Ctrl+3	Toggle palettes
Ctrl+4	To set the sheet palette
Ctrl+6	To set the DB Connect Manager
Ctrl+7	To set the markup set manager palette
Ctrl+8	To open the quick calculator inside the AutoCAD
Ctrl+9	Disable or enable of the command line

1.5.4 SELECTION METHODS IN AUTOCAD:

In AutoCAD, there are several ways to select the objects in a drawing. The following are two primary methods.

Cross Selection:

Drag the cursor from left to right. It selects all objects totally or partially enclosed within the selection rectangle.

Window Selection:

Drag the cursor from right to left. It only chooses things that are totally enclosed within the selection rectangle.

1.5.5 FUNCTIONAL KEYS

The function keys (F1-F12) in AutoCAD can be used to access particular features and carry out a variety of activities.

F1 – It opens AutoCAD help center

F2 – It opens the Auto CAD text window

F3 – It toggles the object snap option

F4 – It toggles the 3Dobject snap option

F5 – It toggles the Isometric drafting mode

F6 – It toggles the dynamic user coordinate system

F7 – It toggles the grid display on and off

F8 – It toggles the ortho mode

F9 – It toggles snap mode

F10 – It toggles polar tracking mode

F11 – It toggles object transparency

F12 – It open the dynamic input toggles object transparency

1.5.6 TYPES OF DRAWING SHEETS

The type of drawing sheet to be selected for a specific purpose depends on the standards of the project, industry norms, and the details required for the construction. The common types of drawing sheets are

(i) Imperial Sheet – It is used to draw the elements of a drawing in inches and feet. The standard dimension of the imperial sheet is 12" X 9".

(ii) Metric Sheet– It is used to draw the elements of a drawing in millimeters and meters.
The standard dimension of the metric sheet is 420mm X 297mm.

1.6 COORDINATE SYSTEMS

A point can be located using two separate coordinate systems in AutoCAD. The X distance (left/right) and the Y distance (up/down) are used in Cartesian coordinate system (or) Rectangular coordinate system. The length (or magnitude) and angle (or direction) between two points are used in the Polar coordinate system. In most cases rectangular coordinates are used when grid positions are known. Similarly, the polar coordinate system is used when the angle and the distance between the points are known.

The three commonly used methods of coordinate systems in Auto CAD

(a) ***Absolute Rectangular Coordinate Method*** – The points of a line are represented as X, Y with respect to origin (0,0).
Syntax: X, Y

(b) ***Relative Rectangular Coordinate Method*** – The coordinates of a point are represented as @ X, Y with respect to the previous point.
Syntax: @± X, ±Y

(c) ***Relative Polar Coordinate Method*** – The coordinates of a point are represented as @ X/ Y < angle in degrees with respect to the previous point.
Syntax: @X distance (or) Y distance < angle

1.7 REFERENCE PLANES/ STANDARD PLANES

Reference planes are essential for understanding and navigating the AutoCAD workspace, particularly when switching from 2D drawing to 3D modelling.

(a) XY Plane (or) Top View: In AutoCAD 2D work XY plane is the default work plane. It is also referred to as the top view with Z-axis constant during 3D modeling work.
(b) YZ Plane (or) Front View: It is also referred to as front view with X-axis constant during 3D modeling work. It is also called as elevation of an object.
(c) ZX Plane (or) Side View: It is also referred to as a side view with the Y-axis constant during 3D modeling work.

1.8 DRAWING USING AUTOCAD

For drawing any plan or object in CAD, the following are the steps involved

1. Fixing the units and limits of the drawing
2. Setting up of required drawing aids
3. Drawing basic entities
4. Using modify tools for fine tuning of the drawing
5. Text and dimensioning
6. Use of Layers and Blocks

Note: Press the ENTER Button after completion of each command line.

1.8.1 INITIAL SETTINGS

COMMAND: **UNITS** (OR) **U**

It is always necessary to clearly specify in which units an object is drawn. The drawing units can be set as mm, meters, inches, feet, degrees, and radians. The length type of the object can be specified as Architecture, Decimal, Engineering, Fractional, Scientific using the units dialogue box shown below.

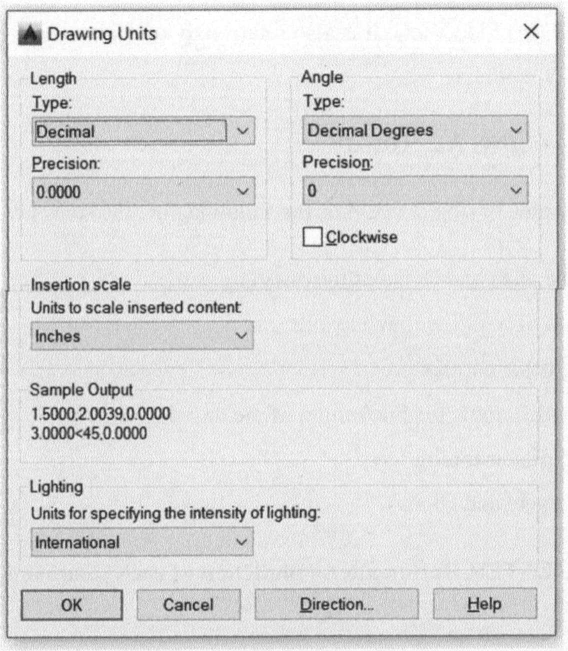

Fig.1.2: Units dialogue box

COMMAND: **LIMITS**

The limit command is used to specify the minimum and maximum limits of the drawing area in which the object has to be drawn. The area beyond these limits cannot be displayed, plotted, or printed.

Command: **LIMITS**

Specify lower left corner or [ON/OFF] <0.0000,0.0000> : 0,0

Specify upper right corner <12.0000,9.0000> : 100,100

COMMAND: **ZOOM (OR) Z**

The Zoom command is used to enlarge or reduce the complete drawing area to view the whole drawing or a specified object clearly in the drawing.

Command: **ZOOM**

Specify the corner of the window, enter a scale factor (nX or nXP), or

[All/Center/Dynamic/Extents/Previous/Scale/Window/Object] <real time>:

Specify the window to zoom by picking the upper right corner and lower left corner by using mouse.

1.8.2 DRAWING AIDS

(i) Grid: An array of dots that are evenly or unevenly spaced along the X and Y axes can be seen on the working screen by using the grid command.

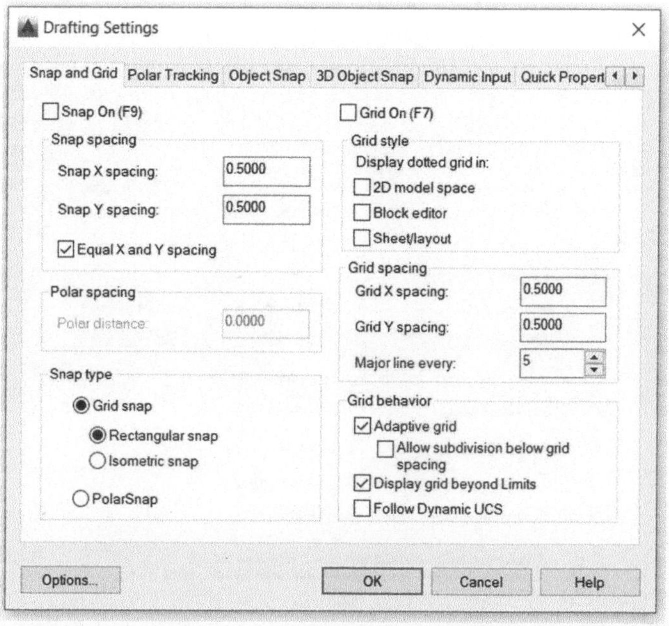

Fig.1.3: Snap and Grid dialogue box

(ii) Snap: Snap mode is specifically used to draw any object on the array of grid displayed.

(iii)Polar Tracking: It allows to you draw a line with specified angle in the grid area.

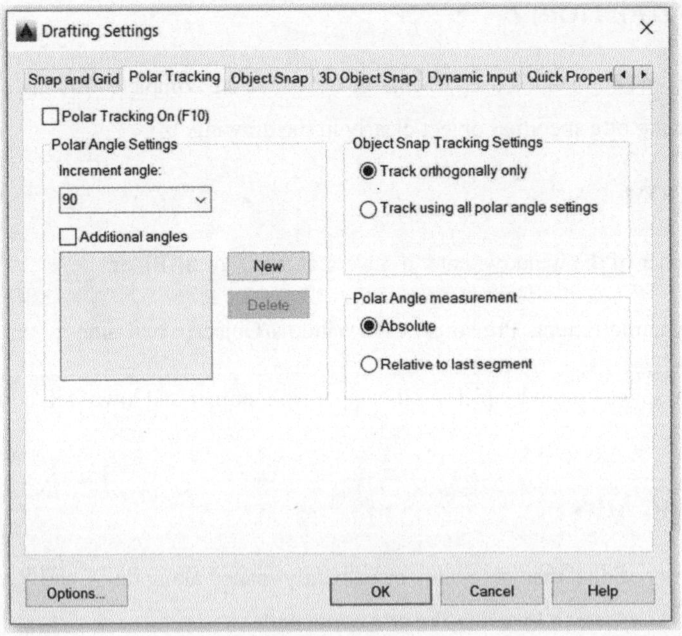

Fig.1.4: Polar Tracking dialogue box

(iv) Object Snap: An important tool used to draw new objects with reference to previous existing elements.

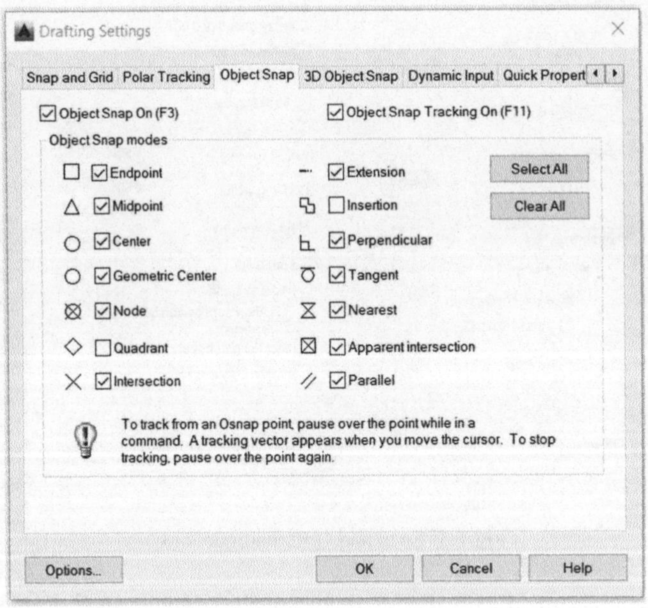

Fig.1.5: Object snap dialogue box

1.8.3 DRAWING BASIC ENTITES

COMMAND: LINE (or) **L**

You can use this command to draw straight lines between particular points. Lines can be drawn by specifying endpoints, lengths and angles. Lines can be drawn in three methods.

(a) Absolute Recatangular Coordinate Method

Command: LINE (or) L

Specify first point: 10,10 or select the point using the mouse

Fig. 1.6: Absolute coordinates

Specify next point or [Undo]: 20,10

(b) Relative Rectangular Coordinate Method

Command: LINE (or) L

Specify first point: 10,10 or select the point using the mouse

Fig. 1.7: Relative coordinates

Specify next point or [Undo]: @10,0

(c) Relative Polar Coordinate Method

Command: LINE (or) L

Specify first point: 10,10 or select the point using the mouse

Fig. 1.8: Polar coordinates

Specify next point or [Undo]: @10<0

To draw the given diagram in Fig 1.9a using three coordinate systems

a)Absolute Rectangular Coordinate Method

COMMAND: **LINE** (OR) **L**

Specify first point : 5,5

Specify next point or [Undo] : 90,5

Specify next point or [exit/Undo] : 90,20

Specify next point or [Close/exit/Undo] : 60,20

Specify next point or [Close/exit/Undo] : 60,35

Specify next point or [Close/exit/Undo] : 30,20

Specify next point or [Close/exit/Undo] : 30,50

Specify next point or [Close/exit/Undo] : 5,50

Specify next point or [Close/exit/Undo] : Press ENTER to complete the drawing

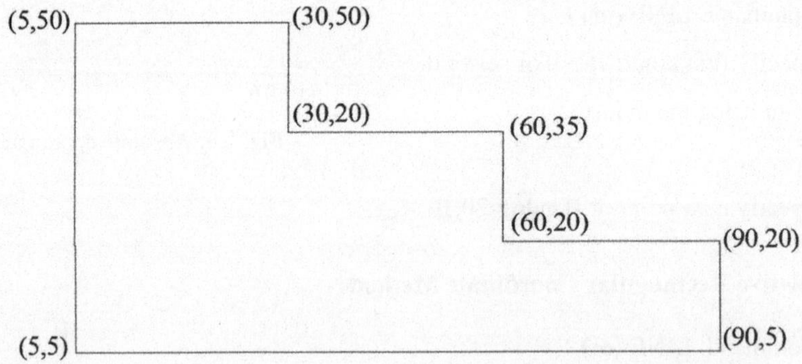

Fig. 1.9 a: Absolute rectangular coordinate method

(b)Relative Rectngular Coordinate Method

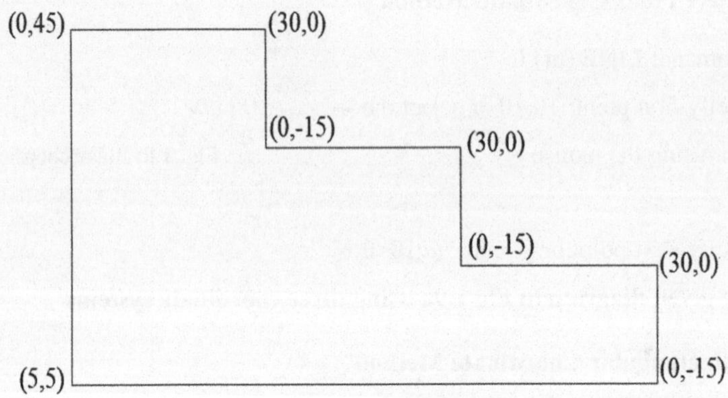

Fig. 1.9 b: Relative rectangular coordinate method

Command: **LINE** (OR) **L**

Specify first point	: 5,5
Specify next point or [Undo]	: 0,45
Specify next point or [exit/Undo]	: 30,0
Specify next point or [Close/exit/Undo]	: 0, -15
Specify next point or [Close/exit/Undo]	: 30,0
Specify next point or [Close/exit/Undo]	: 0, -15
Specify next point or [Close/exit/Undo]	: 30,0
Specify next point or [Close/exit/Undo]	: 0, -15
Specify next point or [Close/exit/Undo]	: -90,0
Specify next point or [Close/exit/Undo]	: Press ENTER to complete the drawing

(c)Relative Polar Coordinate Method

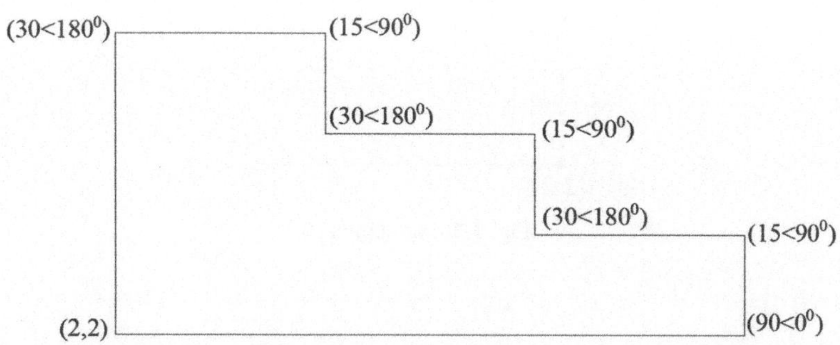

Fig. 1.9 c: Relative polar coordinate method

COMMAND: **LINE** (OR) **L**

Specify first point point by using the mouse	: 2,2 or select a
Specify next point or [Undo]	: 90<360
Specify next point or [exit/Undo]	: 15<90
Specify next point or [Close/exit/Undo]	: 30<180

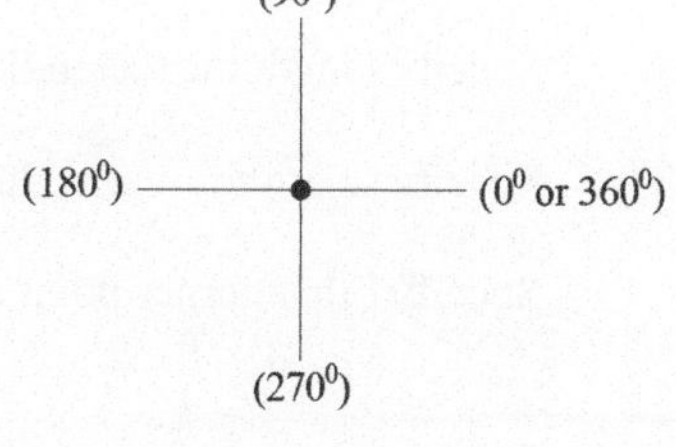

Specify next point or [Close/exit/Undo] : 15<90

Specify next point or [Close/exit/Undo] : 30<180

Specify next point or [Close/exit/Undo] : 15<90

Specify next point or [Close/exit/Undo] : 30<180

Specify next point or [Close/exit/Undo] : 45<270

Specify next point or [Close/exit/Undo]: Press ENTER to complete the drawing.

COMMAND: **RECTANGLE** (or) **REC**

This command allows you to draw the rectangular shapes by specifying two opposite corners. You can use dynamic input or manually enter lengths and widths.

Specify first corner point or [Chamfer/Elevation/Fillet/Thickness/Width] : 2,2 or select any point in the drawing area

Specify other corner point or [Area/Dimensions/Rotation] : @ 10,15

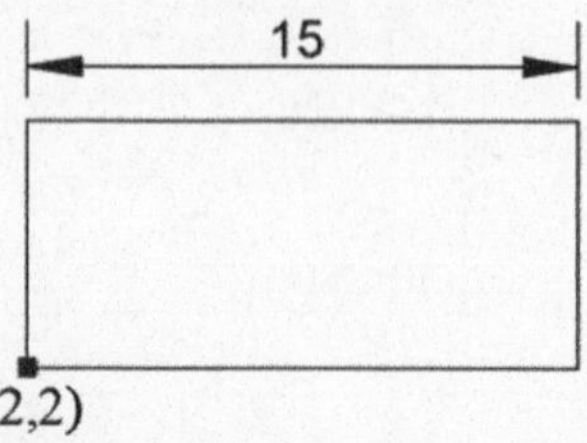

Fig. 1.10: Rectangle

COMMAND: RECTANGLE (or) **REC** (Fig 1.10.)

Specify first corner point or [Chamfer/Elevation/Fillet/Thickness/Width] : 2,2 or select a point by using the mouse

Specify other corner point or [Area/Dimensions/Rotation] : D

Specify length for rectangles <15.0000> : 15

Specify width for rectangles <6.6700> : 10

Specify other corner point or [Area/Dimensions/Rotation] : @15,10

COMMAND: POLY**LINE** (or) **PL**

It creates a 2D polyline

Specify start point: select any point on the screen using mouse.

Current line-width is 0.0000

Specify next point or [Arc/Halfwidth/Length/Undo/Width] : @1<25

Specify next point or [Arc/Close/Halfwidth/Length/Undo/Width] : @1.5<315

Specify next point or [Arc/Close/Halfwidth/Length/Undo/Width] : @1<45

Specify next point or [Arc/Close/Halfwidth/Length/Undo/Width] : @0.5<315

Specify next point or [Arc/Close/Halfwidth/Length/Undo/Width] : Press ENTER to complete the drawing.

Fig. 1.11: Polyline

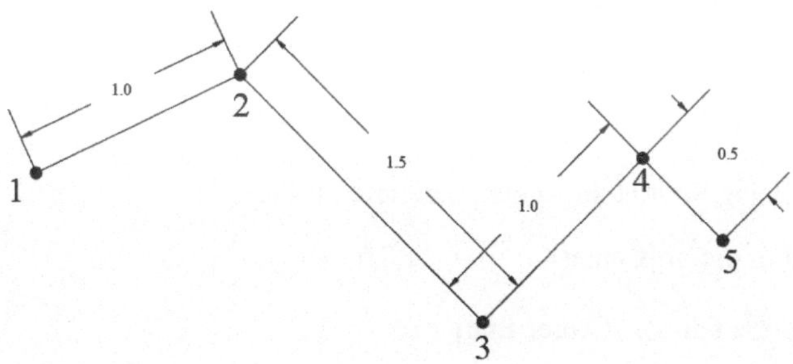

COMMAND: **CIRCLE** (OR) **C**

It creates a circle using a center point and a radius

Drawing a circle by using the radius in Fig 1.12

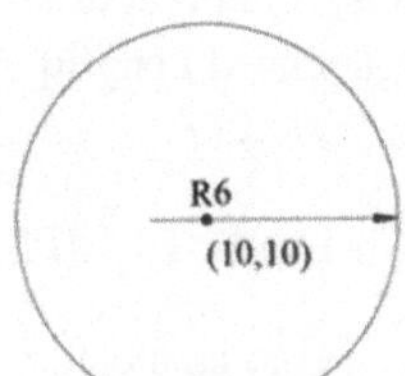

Fig.1.12: Circle

Specify center point for circle or [3P/2P/Tan (tan tan radius)]

: 10,10 or select a center point by using the mouse

Specify the radius of circle or [Diameter] : 6

COMMAND: CIRCLE (or) C

Drawing a circle by using the diameter in Fig1.13

Specify the center point for the circle or [3P/2P/Tan (tan tan radius)]: 10,10 or select a point by using the mouse

Specify radius of circle or [Diameter] : D

Specify diameter of circle :4

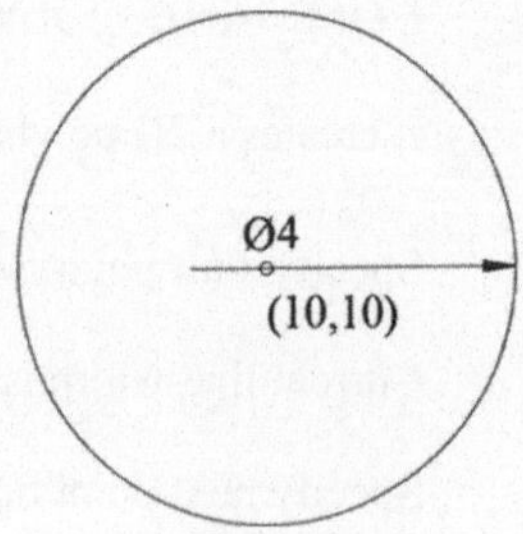

Fig.1.13

COMMAND: **ARC** (OR) **A**

It creates arc

Drawing an arc by using three points in Fig 1.14

Specify start point of arc or [Center] : 5,2.5

Specify second point of arc or [Center/End] : 5

Specify end point of arc : 0,2.5

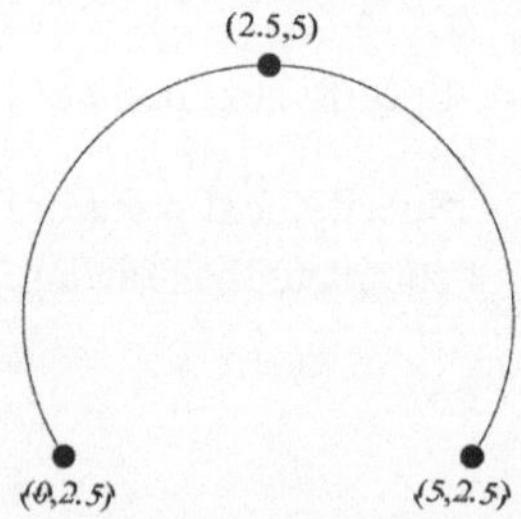

Fig.1.14

COMMAND: **ARC** (OR) **A**

Drawing a arc by using Start point, center, end point in Fig 1.15

Specify start point of arc or [Center] : 5,2.5

Specify second point of arc or [Center/End] : C

Specify center point of arc : @2.5,2.5

Specify end point of arc (hold Ctrl to switch direction) or [Angle/chord Length] : @0,2.5

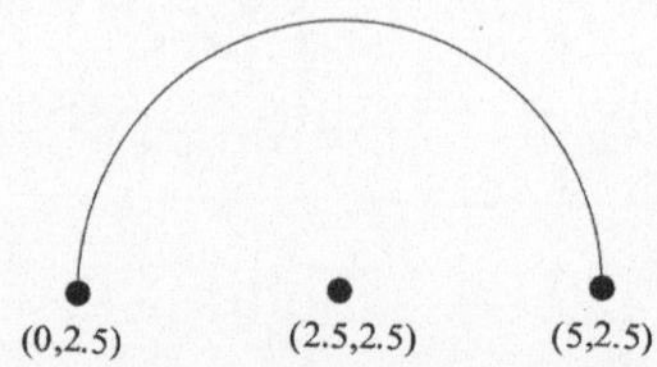

Fig.1.15

COMMAND: **ELLIPSE** (OR) **EL**

It is used to create ellipse or elliptical arcs

Centre - Axis

Specify axis endpoint of ellipse or [Arc/Center] : C

Specify center of ellipse : 2,2

Specify endpoint of axis : 2,4

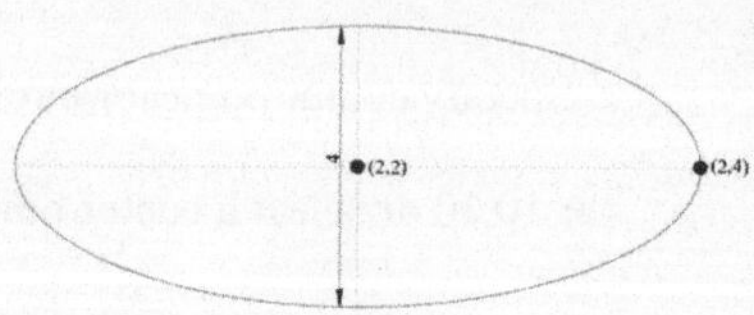

Fig.1.16

Specify distance to other axis or [Rotation] : 5

COMMAND: POLYGON (or) **POL**

It creates an equilateral regular closed polygon in Fig 1.17

Enter the number of sides <4> : 6

Specify center of polygon or [Edge] : E

Specify first end point of edge : Select first
end point on the screen by sing the mouse

Specify second end point of edge : 5

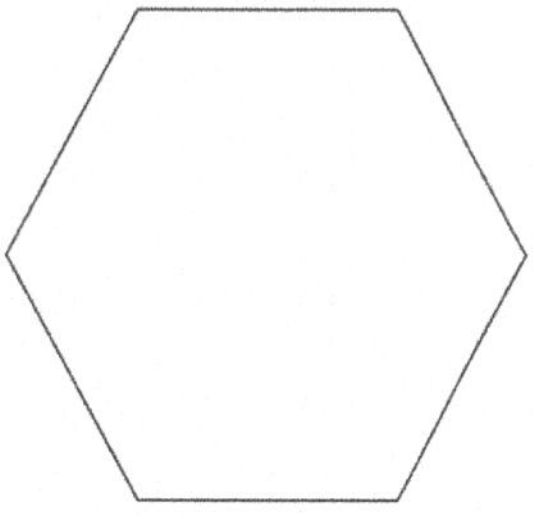

Fig.1.17

COMMAND: POLYGON (or) **POL**

Inscribed in circle in Fig 1.18

Enter the number of sides <4> : 6

Specify center of polygon or [Edge] : Specify center
of the circle using mouse

Enter an option [Inscribed in circle/Circumscribed about circle]
<I>: I

Specify radius of circle : 30

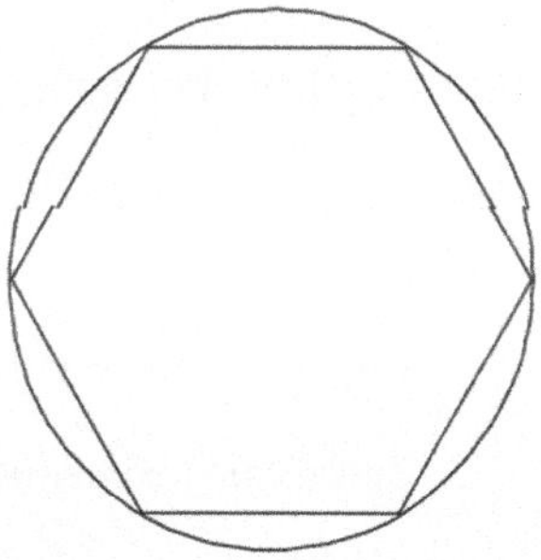

Fig.1.18

COMMAND: POLYGON (or) **POL**

Circumscribed about circle in Fig 1.19

Enter the number of sides <4> : 6

Specify center of polygon or [Edge] : Specify center
of the circle using mouse

Enter an option [Inscribed in circle/Circumscribed about circle] <I>:
C

Specify radius of circle : 30

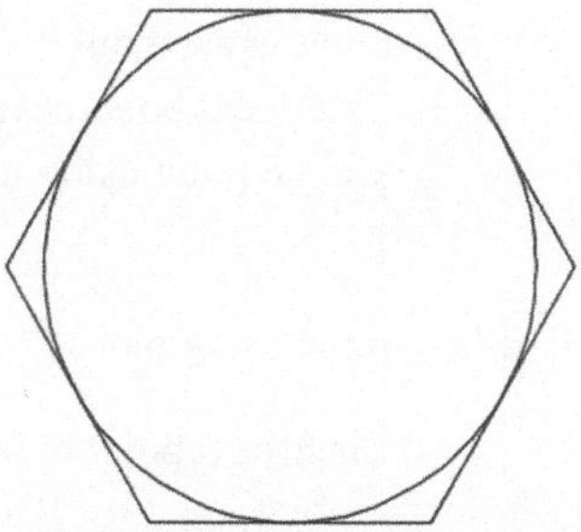

Fig.1.19

COMMAND: CONSTRUCTION LINE (or) **XLINE**

It creates a line of infinite length

Specify a point or [Hor/Ver/Ang/Bisect/Offset]: H

Fig.1.20

Specify through point:

Specify through point:

Specify through point: (enter)

COMMAND: REVISION CLOUD (or) **REVCLOUD**

It creates revision cloud, which is a freehand shape

Specify first point or [Arc length/Object/Rectangular/Polygonal/Freehand/Style/Modify]:A

Specify minimum length of arc: 2

Specify maximum length of arc 4

Specify first point or [Arc length/Object/Rectangular/Polygonal/Freehand/Style/Modify] : R

Specify first corner point or [Arclength/Object/Rectangular]

Fig.1.21

Specify opposite corner:

COMMAND: SPECIAL LINE (or) SPLINE

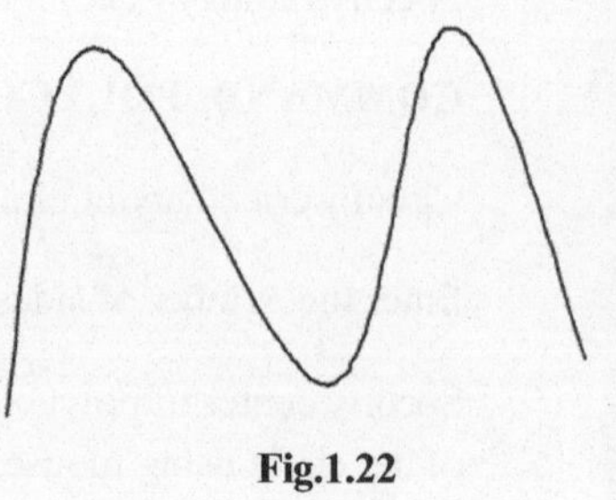

Specify first point or [Method/Knots/Object] : Select first point using mouse

Enter next point or [start Tangency/tolerance] : Select second point using mouse

Enter next point or [end Tangency/tolerance/Undo] : Select next point using mouse

Fig.1.22

Enter next point or [end Tangency/tolerance/Undo/Close]: Select next point using mouse

Enter next point or [end Tangency/tolerance/Undo/Close]: Select end point using mouse

1.8.4 USING MODIFY TOOLS FOR FINE TUNING OF THE DRAWING

COMMAND: COPY (or) **CO**

It is used to copy the objects

Select objects : Select the objects by using mouse

Specify base point or [Displacement/mode] <Displacement>: Select any point as a base point to copy the object

Specify second point or [Array] <use first point as displacement>: Select second point using mouse

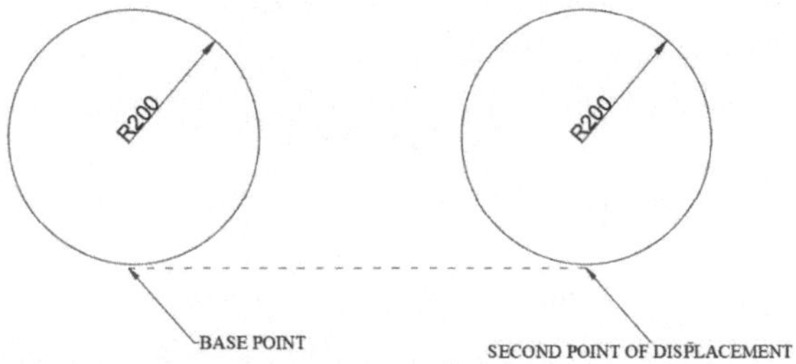

Fig.1.23: Copy

COMMAND: **MOVE** (or) **M**

It is used to move the object from one place to another place

Select objects : Select the objects by using mouse

Specify base point or [Displacement] <Displacement>: Select any point as a base point to move the object

Specify second point or <use first point as displacement>: Select second point using mouse

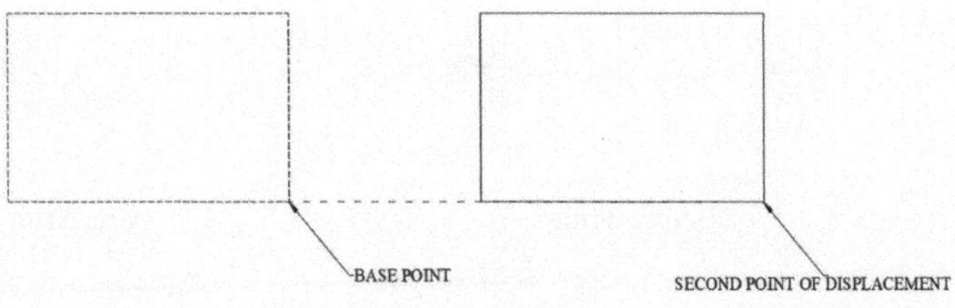

Fig.1.24: Move

COMMAND: OFFSET (or) O

It is used to create parallel lines, concentric circles and parallel curves to the specified distance

Specify offset distance or [Through/Erase/Layer] <0.000> : 200

Select object to offset or [Exit/Undo] <Exit> : Select the object using mouse

Specify point on side to offset or [Exit/Multiple/Undo] <Exit>: Specify the side for offsetting using mouse

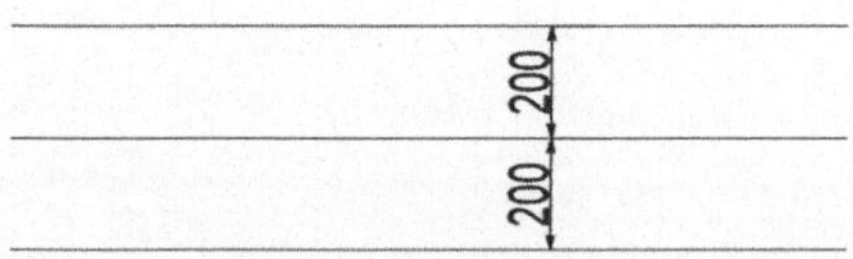

Fig. 1.25: Offset

COMMAND: TRIM (OR) TR

It is used to trim objects to meet the edges of other objects

Select objects or <select all>: Select the objects by using mouse

Select an object to trim or shift-select to extend or

[Fence/Crossing/Project/Edge/erase]: press enter

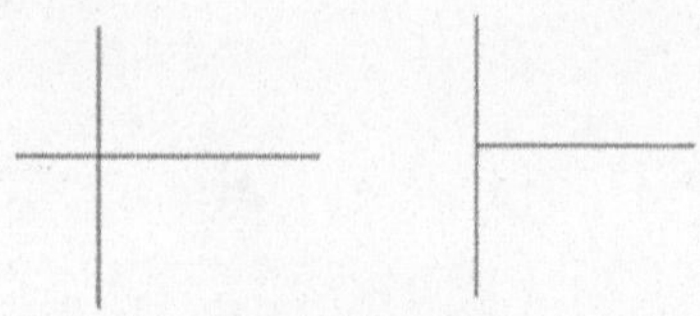

Fig. 1.26: Trim

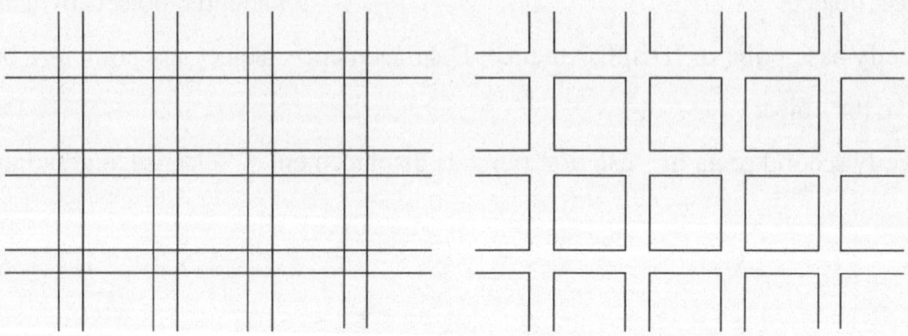

Before Trim After Trim

COMMAND: ROTATE (OR) RO

Rotates objects around a base point

Select objects : Select the object by using mouse

Specify base point : Select any point as a base point

Specify rotation angle or [Copy/Reference] <0>: 30

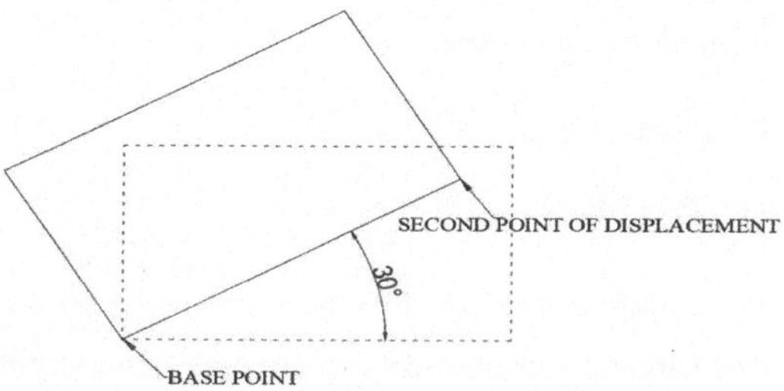

Fig. 1.27: Rotate

COMMAND: SCALE (OR) SC

It enlarges or reduces the selected objects

Select objects : Select the object by using mouse

Specify base point : Select any point as a base point

Specify scale factor or [Copy/Reference] : 2

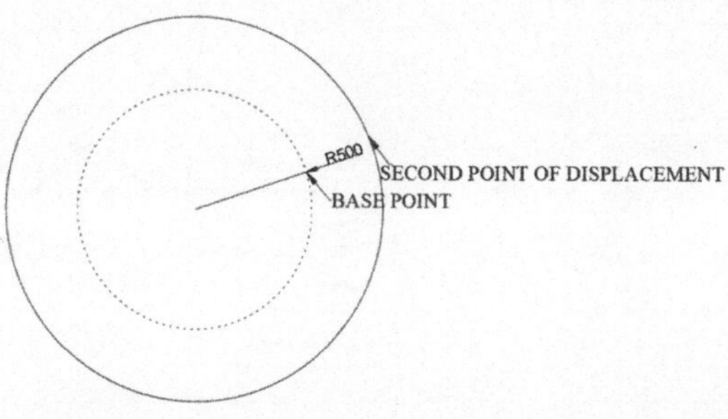

Fig. 1.28: Scale

COMMAND: MIRROR (OR) MI

It creates a mirrored copy of selected objects

Select Object: Select the left side arc using mouse

Specify first point of mirror line: Select center point of the rectangle at top along the Y-axis

Specify second point of mirror line: Select center point of the rectangle at bottom along the Y-axis

Erase source objects? [Yes/No] <No>: N

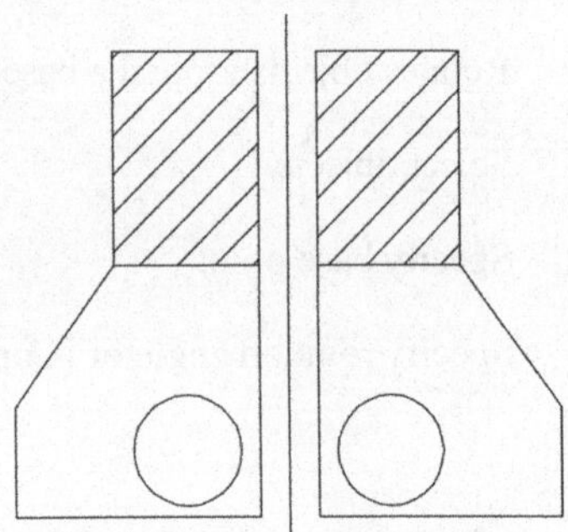

Fig. 1.29: Mirror

COMMAND: HATCH (OR) H

The Hatch command in AutoCAD is used to fill in portions of a drawing with gradients, solid colors, or patterns. In technical and architectural drawings hatching is a basic drafting technique used to visually identify different materials and elements.

Pick internal point or [Select objects/remove Boundaries]: Select the internal point using mouse

Select the type and pattern from the dialogue box as shown below.

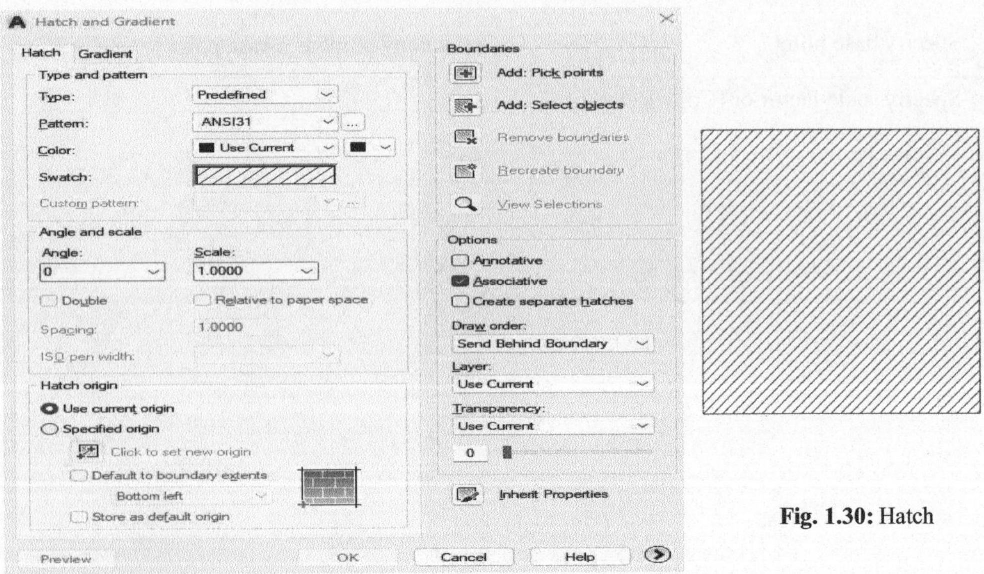

Fig. 1.30: Hatch

COMMAND: GRADIENT HATCH (OR) H

Pick internal point or [Select objects/remove Boundaries]: Select the internal point using mouse

Select the type and pattern from the dialogue box

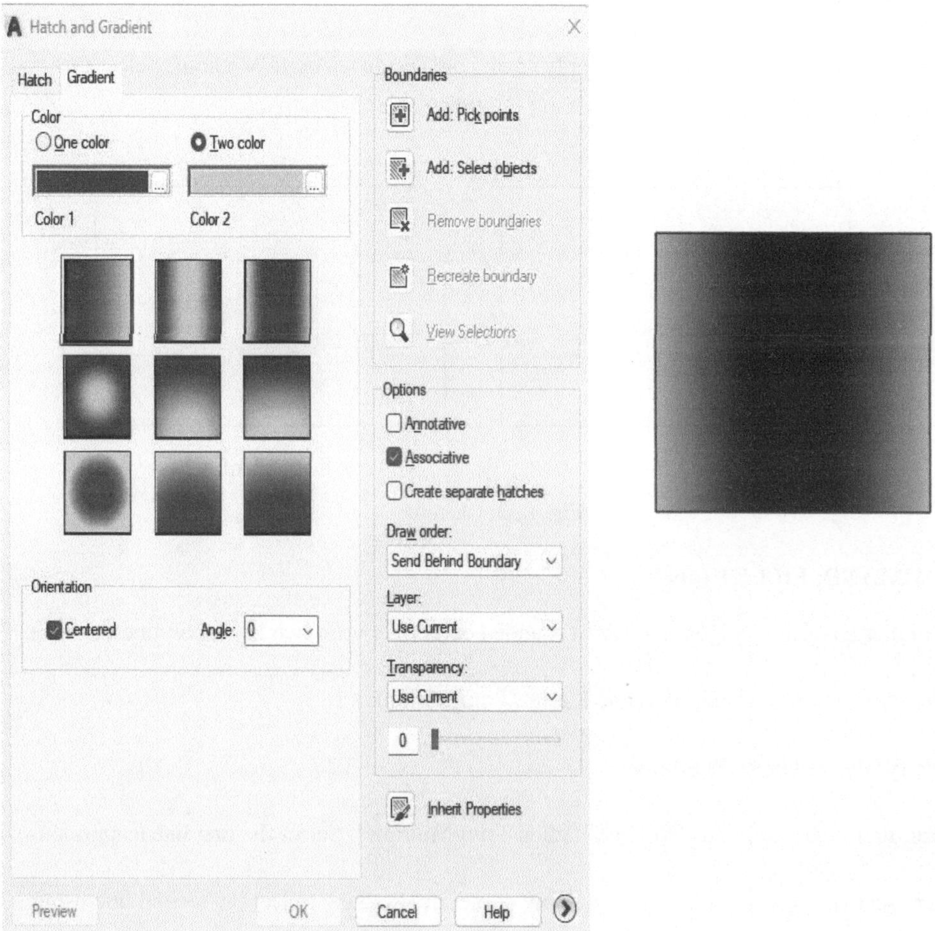

Fig. 1.31: Gradient Hatch

COMMAND: CHAMFER (or) CHA

The Chamfer command is used to create a bevelled edge between two lines or polyline segments.

Select first line or [Undo/Polyline/Distance/Angle/Trim/mEthod/Multiple]: D

Specify first chamfer distance <0.0000> : 1

Specify second chamfer distance <1.0000> : 1

Select first line or [Undo/Polyline/Distance/Angle/Trim/mEthod/Multiple]: Select the first line using mouse

Select second line or shift-select to apply corner or [Distance/Angle/Method]: Select the second line using mouse

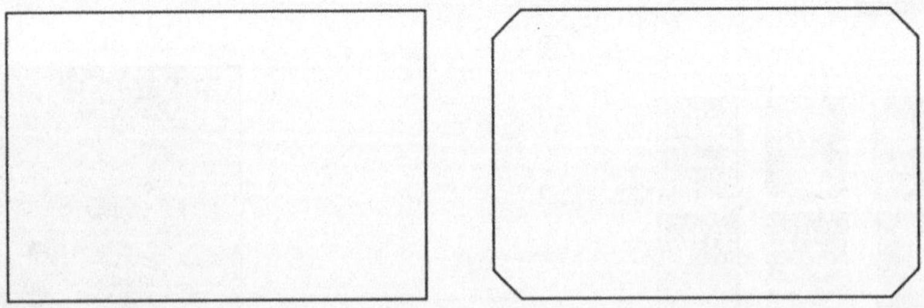

Before Chamfer After Chamfer

Fig. 1.32: Chamfer

COMMAND: FILLET (or) F

The Fillet command is used to create a rounded corner between two lines, polylines, or arcs.

Select first object or [Undo/Polyline/Radius/Trim/Multiple]: r

Specify fillet radius <0.0000> : 2

Select first object or [Undo/Polyline/Radius/Trim/Multiple]: Select the first line using mouse

Select second object or shift-select to apply corner or [Radius]: Select the second line using mouse

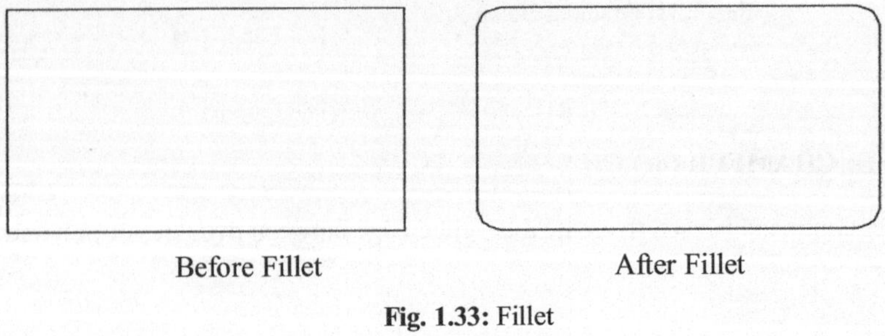

Before Fillet After Fillet

Fig. 1.33: Fillet

COMMAND: ARRAY (or) AR

The array command in AutoCAD is a useful tool for making many duplicates of an object in a pattern. Arrays can be rectangular pattern and polar pattern.

RECTANGULAR ARRAY

Select objects : Select the object using mouse

Select objects : Press ENTER to complete the selection

Enter array type [Rectangular/Path/Polar] <Rectangular> : r

Enter the number of columns or [Expression] <4> :3

Enter the number of rows or [Expression] <3> : 4

Select grip to edit array or [Associative/Base point/Count/Spacing/Columns/Rows/Levels/exit] <exit>: x

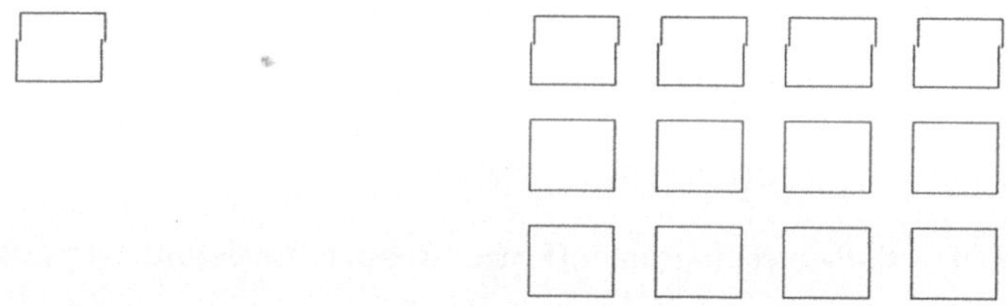

Before Rectangular Array After Rectangular Array

Fig. 1.34: Rectangular array

POLAR ARRAY

Select objects : Select objects in a window using the mouse

Select objects : Press ENTER to complete the selection

Enter array type [Rectangular/Path/Polar] <Polar>: PO

Specify the center point of an array or [Base point/Axis of rotation]: Select the center point of an array using mouse

Enter the no of items or angle between the object: 6

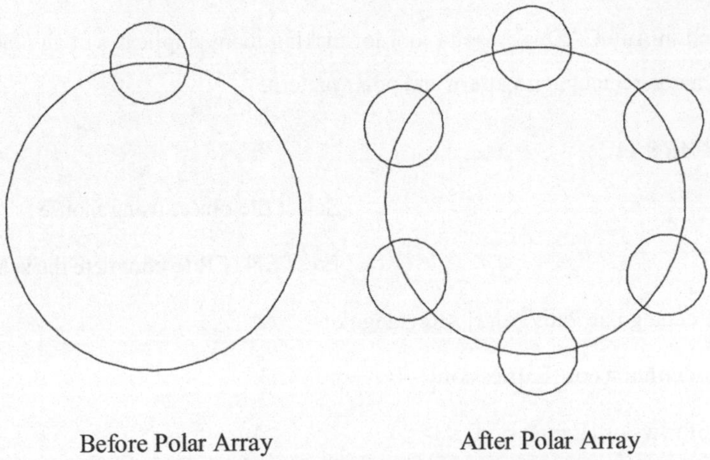

Fig. 1.35: Polar array

COMMAND: EXTEND (or) EX

The Extend is used to extend an object to meet the edges of another object.

Select objects or <select all> : Select the object using mouse

Select object to extend or shift-select to trim or [Fence/Crossing/Project/Edge/Undo]: Select the object using mouse

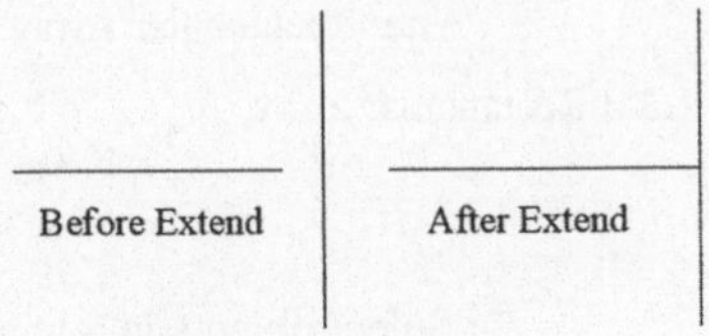

Fig. 1.36: Extend

COMMAND: EXPLODE

The explode command is used to divide complex things into simpler components.

Select Object : Select the object using mouse

Press ENTER

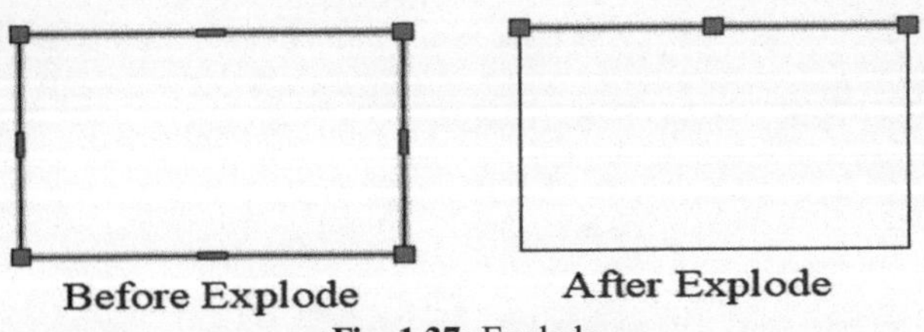

Fig. 1.37: Explode

COMMAND: BREAK (or) BR

The Break command allows you to delete a part of an item or create a gap.

Select object : Select the first break point using mouse

Specify second break point or [First point] : Select the second break point using mouse

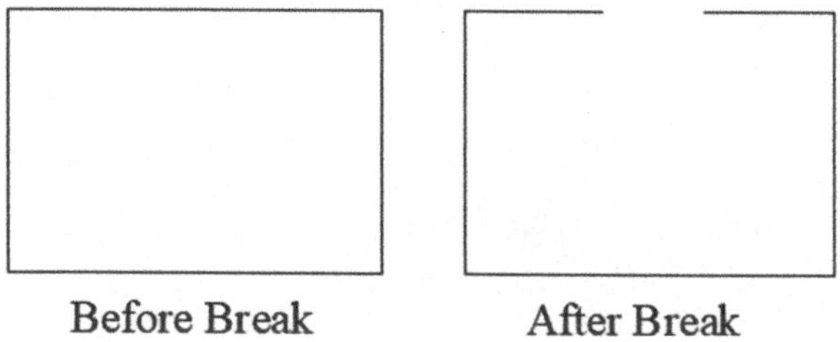

Fig. 1.38: Break

COMMAND: BREAK LINE

The Break Line is used to signify where a section of the drawing has been broken or shortened.

Fig. 1.39: Breakline

COMMAND: JOIN (or) J

The joint command is used to combining many items into a single object.

Select source object or multiple objects to join at once : Select the object using mouse

Select objects to join : Select the object using mouse

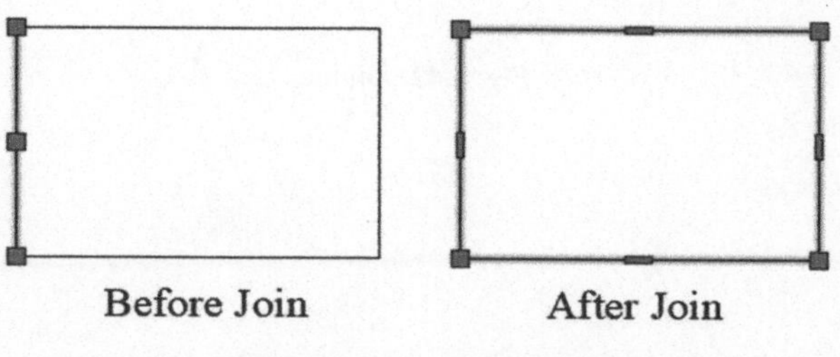

Fig. 1.40: Join

COMMAND: POINT (or) PO

The Point command is used to define locations where objects will be placed or measurements made, mark specific coordinates, or acts as a reference for other objects. It represents a single position in the drawing area.

Specify a Point : Specify a point using mouse

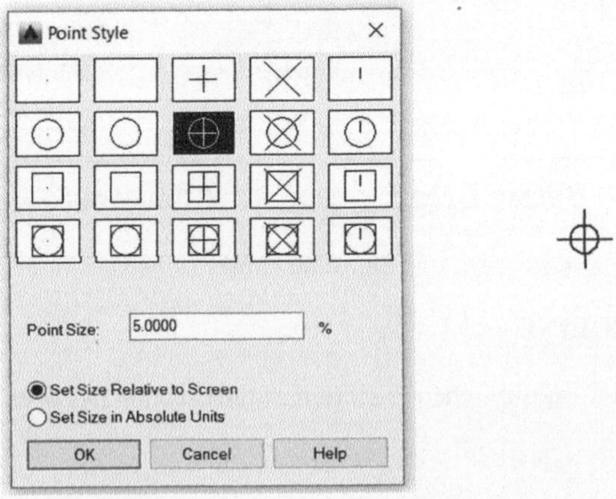

Fig. 1.41: Pointstyle dialogue box

COMMAND: DIVIDE (or) DIV

The Divide command is used to set a specified number of evenly spaced point objects along the length of an object such as a line, polyline, arc, circle or ellipse.

Select object to divide : Select the object using mouse

Enter the number of segments or [Block] : 4

Fig. 1.42: Divide

COMMAND: ID

The ID command displays the coordinate values for a given position in the drawing area.

Specify ID point : Select the point using mouse

```
Command: ID
Specify point:  X = 373.4614     Y = -32.5086      Z = 0.0000
```

Fig. 1.43: ID

COMMAND: DISTANCE (or) DIST

The Distance command is used to measure the distance between the two points.

Specify first point : Select the first point using mouse

Specify second point or [Multiple points] : Select the second point using mouse

```
Distance = 11.9888,  Angle in XY Plane = 0,  Angle from XY Plane = 0
Delta X = 11.9888,  Delta Y = 0.0000,   Delta Z = 0.0000
```

Fig. 1.44: Distance

COMMAND: AREA

The Area command is used to calculate the area and perimeter of a selected object.

Specify first corner point or [Object/Add area/Subtract area] <Object>: Select the first point using mouse

Specify next point or [Arc/Length/Undo] : Select the next point using mouse

Specify next point or [Arc/Length/Undo] : Select the next point using mouse

Specify next point or [Arc/Length/Undo] : Select the next point using mouse

Specify next point or [Arc/Length/Undo/Total] <Total>: Select the next point using mouse

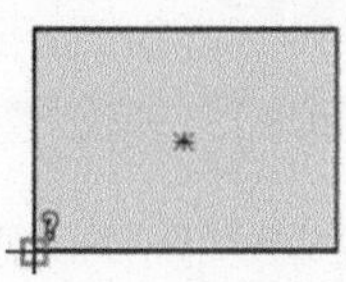

```
Area = 22.0467, Perimeter = 19.0655
```

Fig. 1.45: Area

COMMAND: QUICK CALCULATOR (or) QUICKCALC (or) QC

Quick Calc is a flexible calculator in AutoCAD that may be used to directly use AutoCAD drawing values in calculations, as well as carry out complex mathematical operations and unit conversions in addition to normal arithmetic operations.

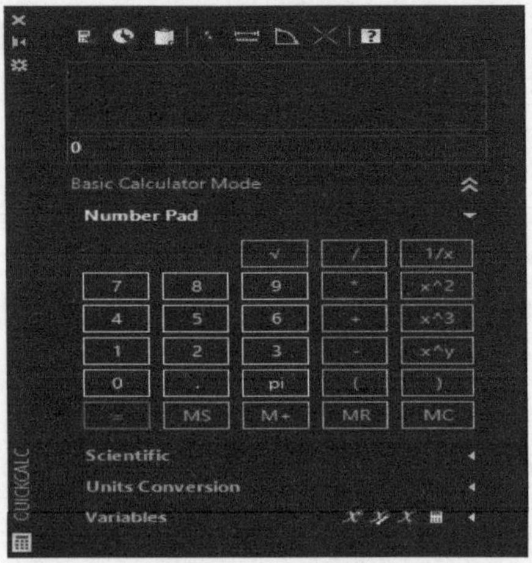

Fig. 1.46: Quick Calculator

1.8.5 TEXT AND DIMENSION

COMMAND: **TEXT** (OR) **T**

Text command is used to create a single line of text in AUTOCAD

Current text style: "Standard" Text height: 1.5000 Annotative: No Justify: Left

Specify start point of text or [Justify/Style]: Specify the first point of text using mouse

Specify height <1.5000> : 2

Specify rotation angle of text <0> : 0

Fig. 1.47: Text

COMMAND: **MULTILINETEXT** (OR) **MTEXT**

The Multiline text is used to create text consisting of multiple lines.

Specify first corner : Specify the first corner using mouse

Specify opposite corner or [Height/Justify/Line spacing/Rotation/Style/Width/Columns]:

Specify opposite corner using mouse

Fig. 1.48: Multi Text

COMMAND: DIMENSION LINEAR (OR) **DIMLIN**

It creates a linear dimension for the object.

Specify first extension line origin or <select object> : Select first point of the line using mouse

Specify second extension line origin : Select second point of the line using mouse

Specify dimension line location or [Mtext/Text/Angle/Horizontal/Vertical/Rotated]: Select the location away from the object

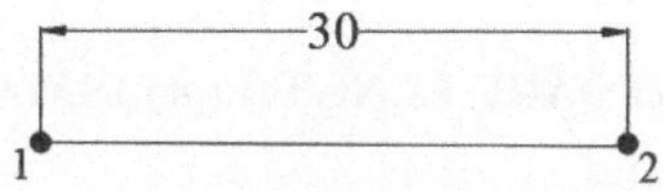

Fig. 1.49: Dimension Linear

COMMAND: DIMEMSION ALIGNED (OR) **DIMALI**

It creates an aligned linear dimension

Specify first extension line origin or <select object>: Select first point of the line using mouse

Specify second extension line origin : Select first point of the line using mouse

Specify dimension line location or [Mtext/Text/Angle]: Select the location away from the object

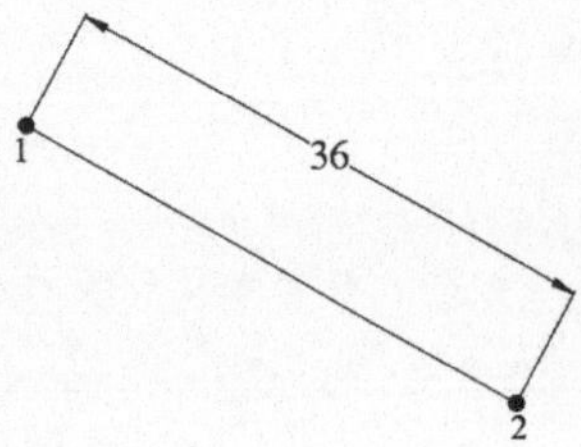

Fig. 1.50: Dimension Aligned

COMMAND: DIMENSION ANGULAR (OR) **DIMANG**

It creates an angle between the two lines, arc, circle

Select arc, circle, line, or <specify vertex> : Select first point of the line using mouse

Select second line : Select first point of the line using mouse

Specify dimension arc line location or [Mtext/Text/Angle/Quadrant]: Specify the location away from the object

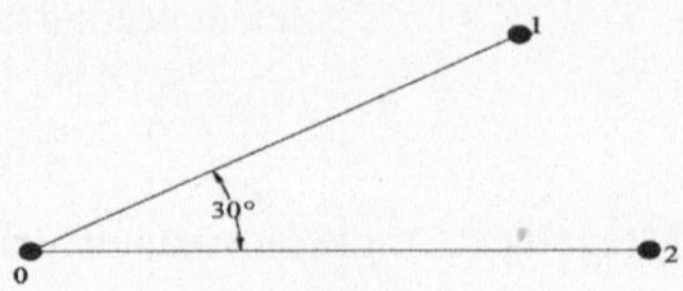

Fig. 1.51: Dimension Angular

COMMAND: DIMENSION ARC LENGTH (or) DIMARC

It creates the length of an arc

Select arc or polyline arc segment : Select the arc or polyline using mouse

Specify arc length dimension location, or [Mtext/Text/Angle/Partial/Leader]: Select the location away from the object

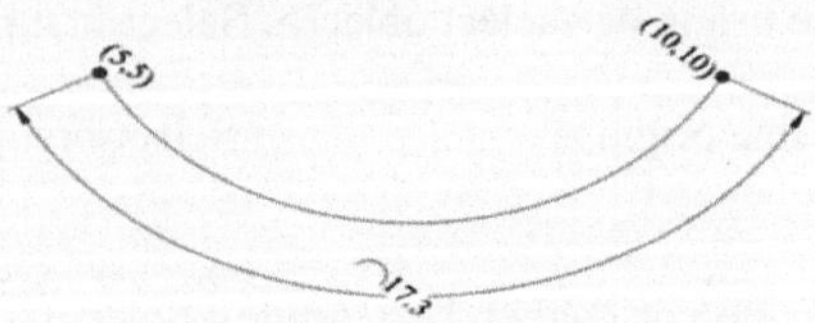

Fig. 1.52: Dimension Arc length

COMMAND: DIMENSION DIAMETER (or) DIMDIAMETER

It creates a diameter for a circle or arc

Select arc or circle : Select circle or arc using mouse

Specify dimension line location or [Mtext/Text/Angle]: Select the location away from the object

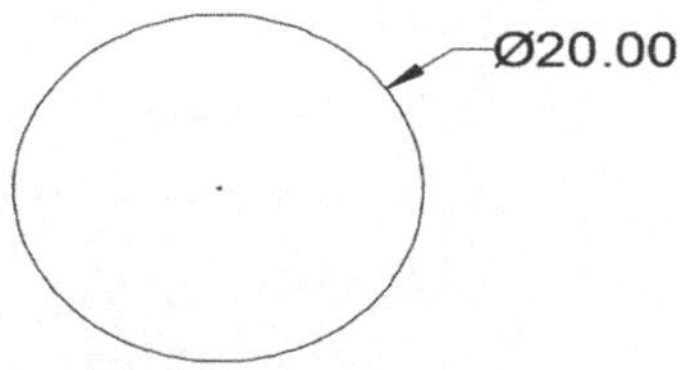

Fig. 1.53: Dimension Diameter

COMMAND: DIMENSIONRADIUS (or) DIMRADIUS

Select arc or circle : Select arc or circle using mouse

Specify dimension line location : Select the location away from the object

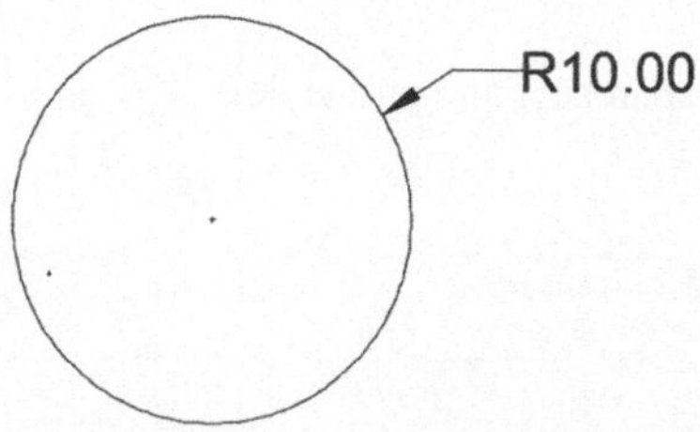

Fig. 1.54: Dimension Radius

COMMAND: MULTILEADER (OR) MLEADER

Specify leader arrowhead location or [leader Landing first/Content first/Options] <Options>:

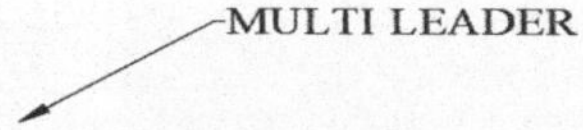

Fig. 1.55: Multileader

Specify leader landing location: Select the location away from the object

1.8.6 LAYERS AND BLOCKS

1.8.6.1 LAYERS

In AutoCAD, layers are a fundamental organizational element that allows you to manage and control the visibility and properties of object in your drawing.

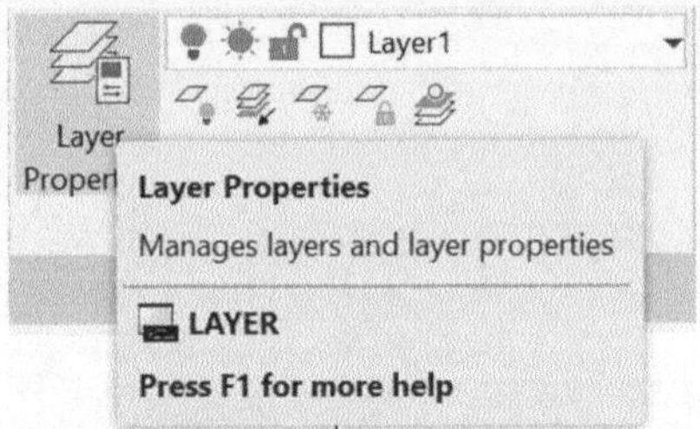

Fig. 1.56: Layer Properties

(i) New layers

The layers properties manager or the layer command can be used to create a new layer

Click the "new layer" button and then enter a name for the layer as well as parameters like colour,

line type, line weight, transparency, and plot style

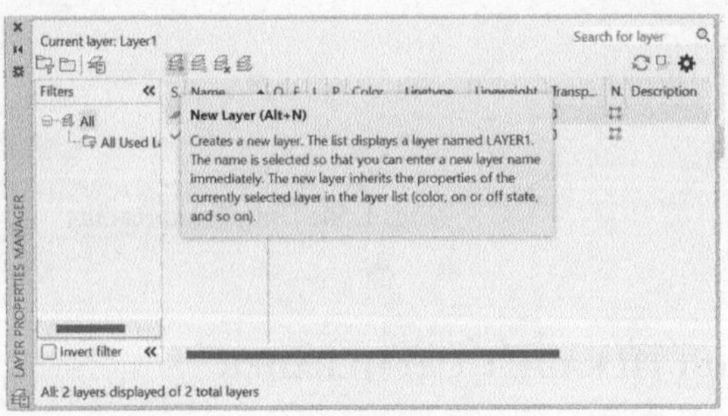

Fig. 1.57: New layer

(ii) Set the current layers

The "current layer" is the layer to which new objects are added. You can set the current layer by selecting it from the layer dropdown on the home tab or by using the 'laymcur' command.

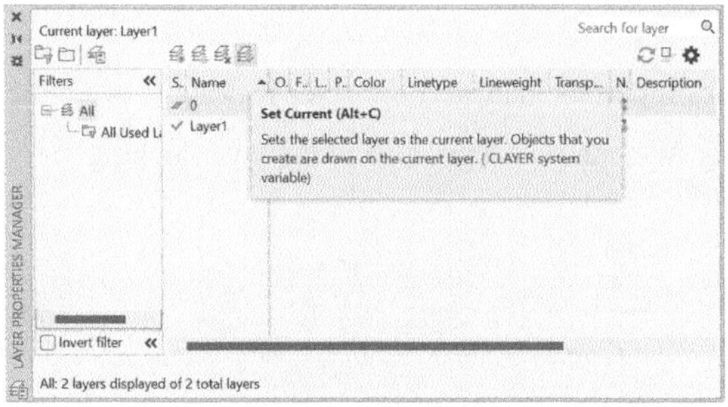

Fig. 1.58: Selection of current layer

(iii) Layers properties manager

The layer properties manager helps you to manage the all-layer aspects. It also provides freeze, thaw, lock, unlock, and change various properties

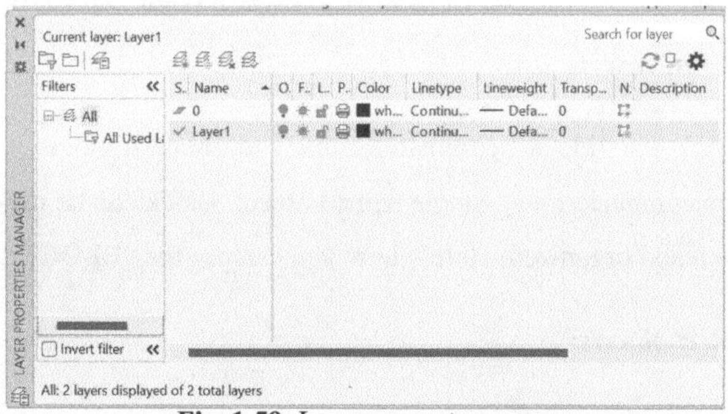

Fig. 1.59: Layer property manager

(iv) Layers properties

Every layer has different properties like colour, line type, line weight, transparency and plot style. You can change these properties using the layer properties manager or double clicking on a layer in the layer in the layer properties palette

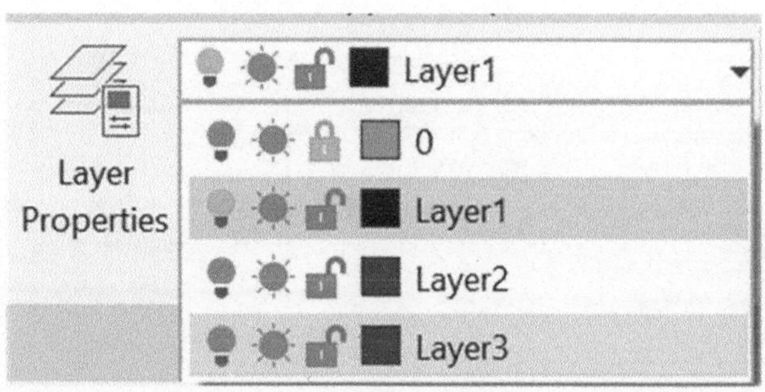

Fig. 1.60: Layer properties

(v) Layers locking

In the layer properties manager, the layer locking is available. It is use to lock layer in the drawing. We can give a pin also to the layer locking drawing.

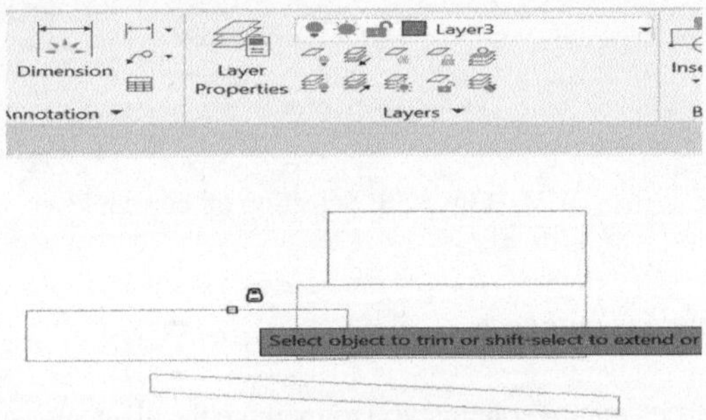

Fig. 1.61: Layer locking

1.8.6.2 Block and Insert

In AutoCAD, the "block" command is used to create a block, which is a collection of objects that are combined into a single named object. Block can be reused in drawing, making it more efficient and organized. Here's how you can use the "BLOCK" command:

(i) Select objects:

Type 'BLOCK' in the command line or use the 'BLOCK' icon in the home tab on the ribbon - Press enter

COMMAND: BLOCK (or) B

Specify insertion base point : Specify the base point using mouse

Select object : Select the object using mouse

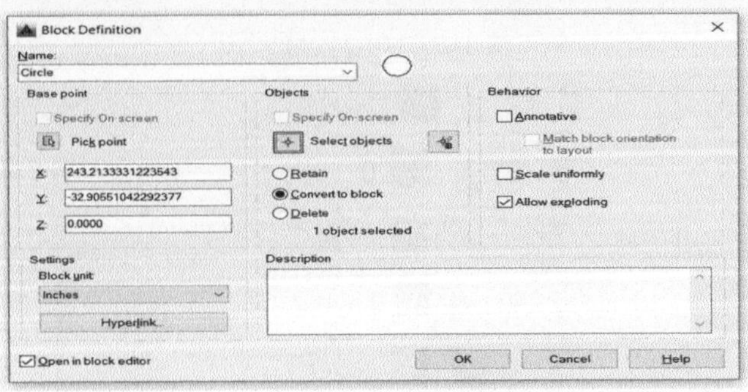

Fig. 1.62: Block

(ii) ***Specify block name:***

Give a name to your block in the name entering box

Choose the object that you are going to make block by selecting

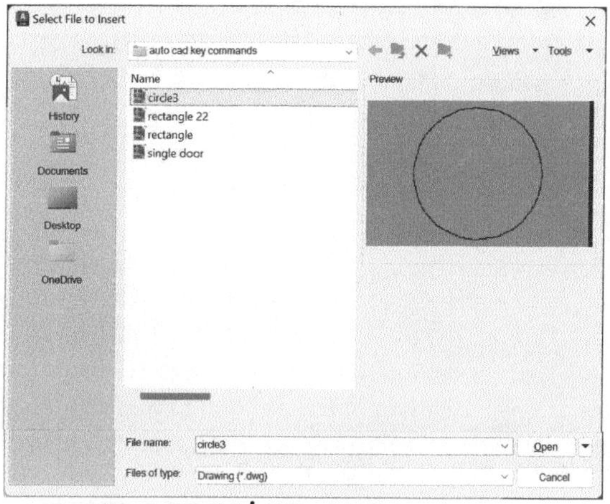

Fig. 1.63: Select file to insert

(iii) ***Insertion point:***

Specify the base point or insertion point for the block. This point is used as the reference point when you insert the block into a drawing.

(iv) ***Other options:***

You can set options like scaling, rotation and more depending on your requirements. These options are available in the block definition dialog box.

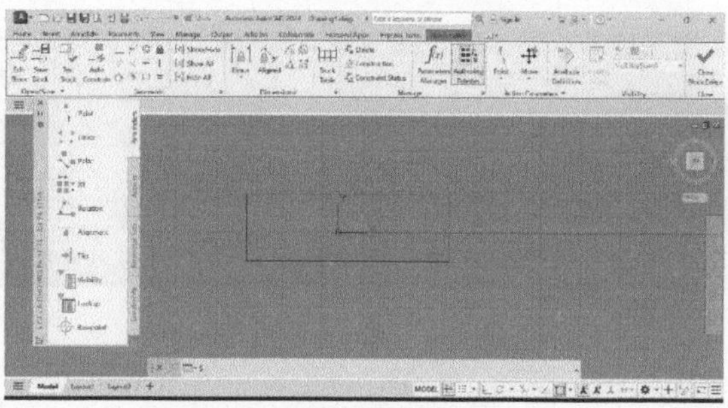

Fig. 1.64: other options

(v) OK:

Once the desired options have set, click "OK" or press enter. The block is now created

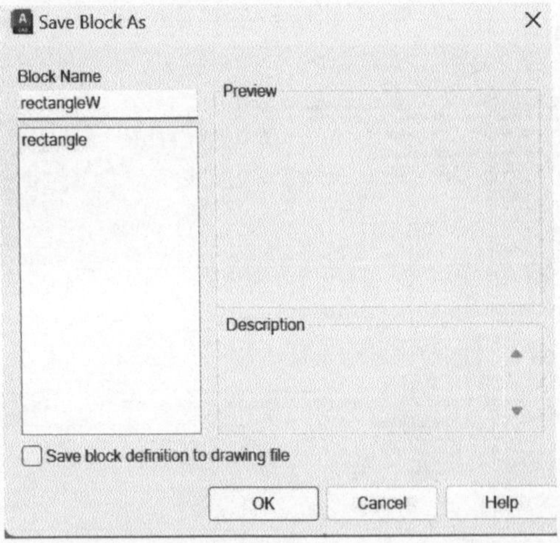

Fig. 1.65: Save block as

To insert a block into drawing after creating it:

- Type 'INSERT' in the command line or use the 'INSERT' icon in the ribbon.
- Select the block from the list or browse to find the drawing file that contains the block

COMMAND: INSERT (or) I

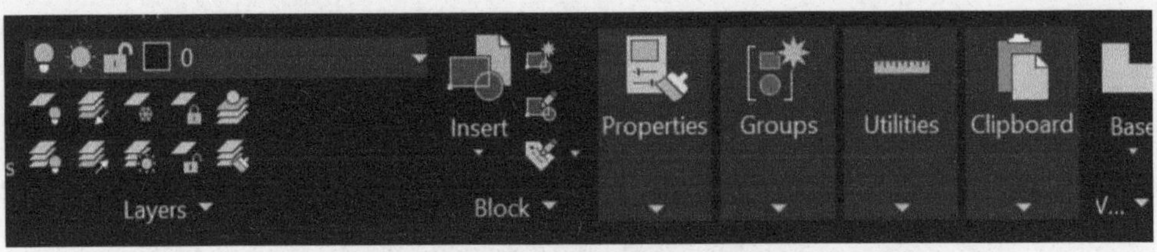

Fig. 1.66: Insert block

Specify Insertion Point: Choose the insertion point for the block in your drawing

Scale/Rotation: The block created can be reduced or scaled up based on the plan dimensions.

OK: Click "OK" or press enter to insert the block.

UNIT SUMMARY

After thoroughly going through the unit, you should remember that

- The Drawing should replicate your idea in a clear and transparent way.
- All the minor and major details in a drawing should be clearly noted and referenced in the drawing with a special note.
- Use of AutoCAD makes your drawing job easier and more efficient than developing a drawing by hand.
- Mistakes/changes can be corrected easily using CAD.
- While defining (or) drawing an object, the tools and the coordinate systems play a vital role.
- All the commands have to be clearly studied before utilizing the specific command.

EXERCISES

A. Multiple Choice Questions

1. The functional key for grid mode on/off ____________
   a) F6
   b) F7
   c) F8
   d) F9
2. What is the shortcut for undo?
   a) Ctrl+Z
   b) Ctrl+Y
   c) Ctrl+9
   d) Ctrl+0
3. Shortcut key for command window ____________
   a) Ctrl+0
   b) Ctrl+1
   c) Ctrl+8
   d) Ctrl+9
4. Users can access AutoCAD commands in ____________
   a) Status bar
   b) Command window
   c) Drawing area
   d) Modify window

5. Chamfer command is used to produce ____________
   a) Sharp corners
   b) Point corners
   c) Round corners
   d) None of the above
6. Shortcut key for polygon ____________
   a) Pol
   b) Pl
   c) Poly
   d) Polygon
7. The file extension that can be accessed by AutoCAD ____________
   a) DOC
   b) DWT
   c) DWG
   d) DXF
8. Shortcut for saving AutoCAD drawings ____________
   a) Shft+S
   b) Alt+S
   c) Ctrl+Alt+S
   d) Ctrl+S
9. The shortcut command for Explode ____________
   a) Ex
   b) Exp
   c) Expl
   d) None of the above
10. Which command is used to obtain parallel lines ____________?
    a) Extend
    b) Trim
    c) Offset
    d) Chamfer

Answers

1. b 2. a 3. d 4. b 5. a 6. a 7. b 8. d 9. d 10. C

B. Short and Long Answer Type Questions

1.1 List out the elements of drawing for showing information?

1.2 Write a short note on user interface with a neat sketch?

1.3 List out the commands in draw tool bar. Explain with examples?

1.4 List out the command in modify tool bar. Explain with examples?

1.5 Discuss the importance of layers in AutoCAD?

KNOW MORE

- *CAD Command Shortcuts:* All the commands listed above have specific shortcuts in CAD that can be directly used on the keyboard without using toolbars (or) command lists. Use of these shortcuts saves your time and energy.
- *Programming:* AutoCAD also supports to developing customized tools (or) commands given by the user with the help of developing a program in the background using "LISP". Auto LISP is an important tool for engineers who have more repetitive work.
- *Import/Export options:* Drawings in AutoCAD can be transferred (or) opened in any other structural (or) architectural software's like STAAD-Pro, REVIT, and Sketchup, etc.
- *Block Diagram:* All the layers/elements of a single drawing can be converted into a single unit. This option enables the engineer's data not to get edited after their drawing finalization.

REFERENCES AND SUGGESTED READINGS

1. T Jeyapoovan. Engineering Drawing and Graphics using AutoCAD. 3rd ed.; 2014.
2. Cadfolks. AutoCAD 2020 For Beginners, 7th ed.; Kishore, 2019.
3. B.V.R Gupta and M. Raja Roy. Engineering Drawing with AutoCAD, 3rd ed.; Wiley Publications, 2020.

Dynamic QR Code for Further Reading

2 SYMBOLS AND SIGN CONVENTIONS

UNIT SPECIFICS

The following aspects are discussed in this module.

- Conventional signs for all types of materials used in construction
- Basics of rebar drawings
- Welding and structural steel drawings
- Dimensioning standards

RATIONALE

The effective construction of a structure is depending on how well the engineer understands the drawings. This unit explains the identification of various materials used in the buildings using signs and symbols mentioned in the drawings. Further, fundamental knowledge of the rebar drawings is issued for a RCC framed structures and welding symbols for structural steel fabrication drawings.

PRE-REQUISITES

- Construction materials
- Structural elements in a framed building
- Connections in steel structures

UNIT OUTCOMES

The following are the outcomes of this unit:

U2-O1: Basic knowledge of signs and symbols for all the materials

U2-O2: Understanding how rebar drawings are issued

U2-O3: Understand how fabrication drawings are issued

U2-O4: Learning the basics of dimensioning

Unit- 2 Outcomes	**EXPECTED MAPPING WITH COURSE OUTCOMES** (1-Weak correlation; 2-Medium correlation; 3-Strong correlation)				
	CO-1	**CO-2**	**CO-3**	**CO-4**	**CO-5**
U2-O1	3	3	2	1	1
U2-O2	1	2	3	1	2
U2-O3	1	3	3	1	2
U2-O4	1	3	2	2	1

INTRODUCTION

The most important tool for understanding the drawings issued by architects, engineers, electricians, and the plumbing division is to understand the sign conventions and symbols used for the material to be used in that particular element. An engineer cannot prepare an abstract estimate for different types of materials required for a project construction without having a basic knowledge of material signs.

Importance of conventional signs

- It is a standard language for architects and engineers to express their ideologies through simple drawings.
- Irrespective of the state and nation, they provide a unified code that can be transferred with minor errors.
- The use of standardized signs and symbols can reduce the effort of communicating rapidly.
- Essential for conveying information quickly, accurately, and ensuring a more efficient work path.

The conventional signs and symbols used for different materials used in the construction industry, architectural symbols, structural, topographical, electrical, and plumbing are shown in the next section of this chapter.

2.1 MATERIALS

In many construction projects, conventional building materials including Portland cement concrete, wood, steel, and asphalt etc. are frequently used. Due to their ease of recognition, those materials must contain symbols and sign conventions. Following are the conventional signs for some standard building materials.

ELEMENT	SYMBOL	ELEMENT	SYMBOL
BRICK	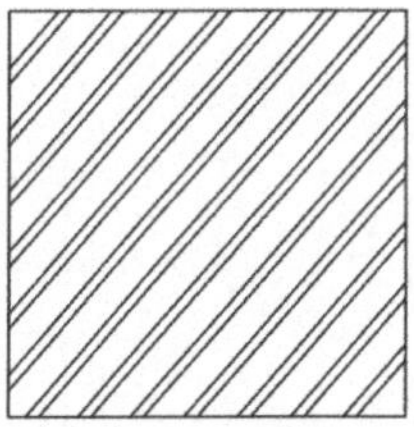	STONE EXISTING	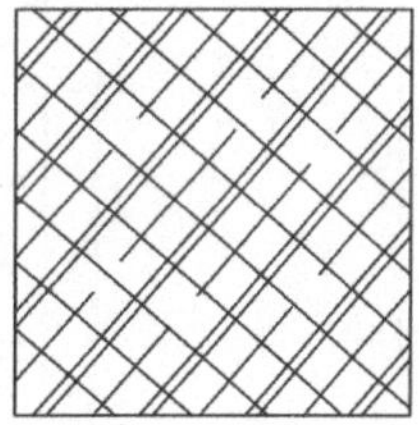
BRICK EXISTING	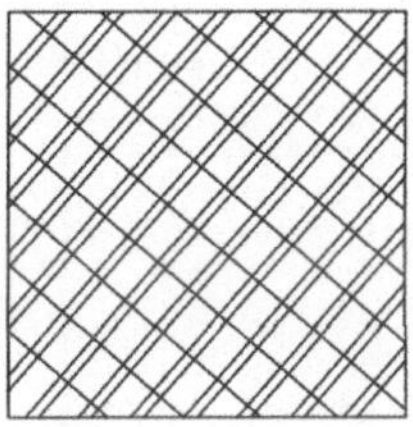	ASHLAR STONE	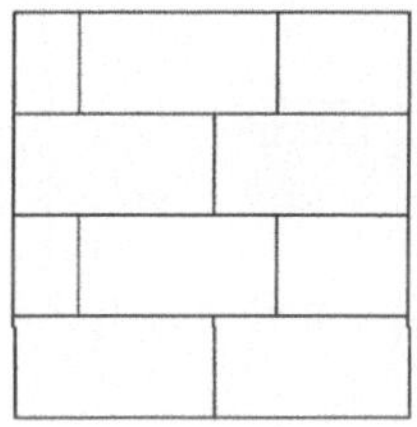
FIRE BRICK	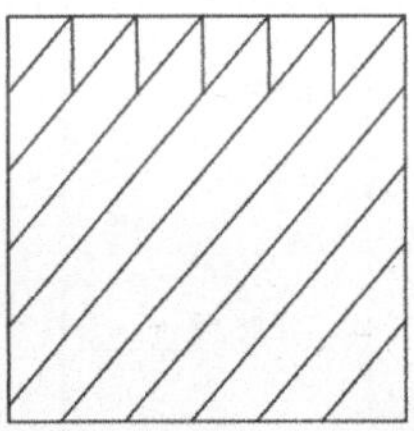	RUBBLE	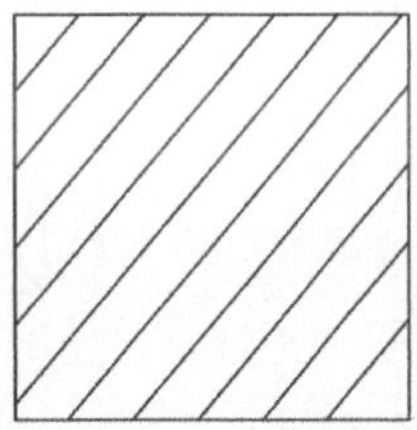
SPECIAL BRICK	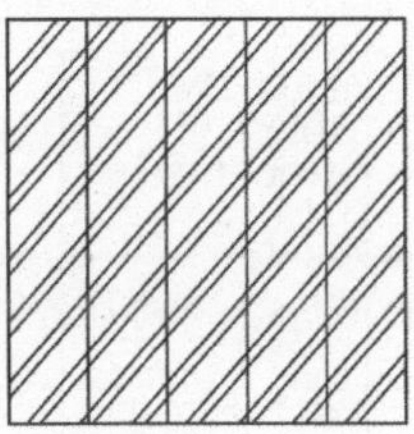	RUBBLE STONE	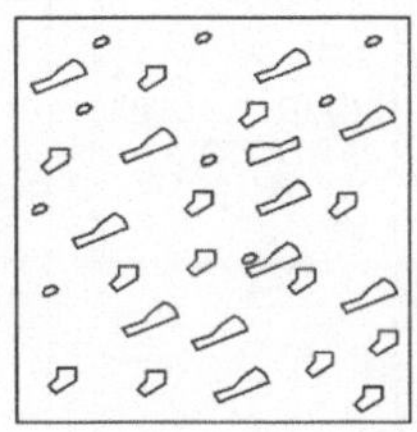
STONE	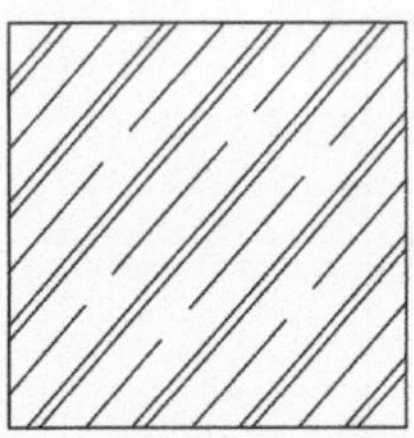	MARBLE SLATE PORCELIN STONEWARE	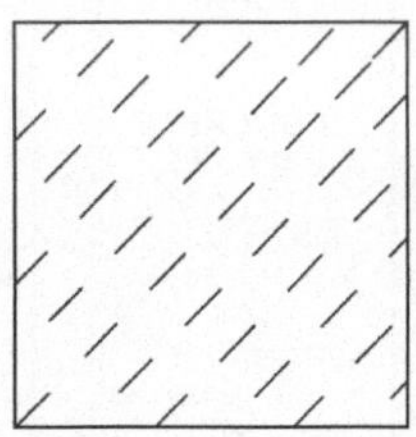

ELEMENT	SYMBOL	ELEMENT	SYMBOL
PLASTER	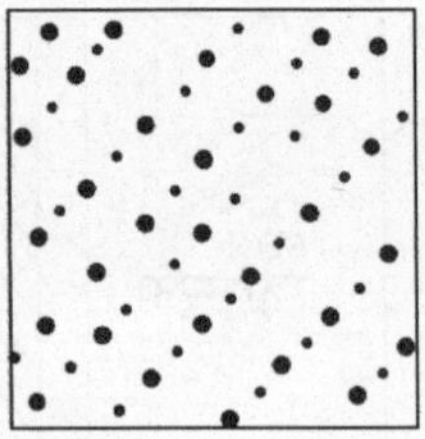	CONCRETE EXISTING	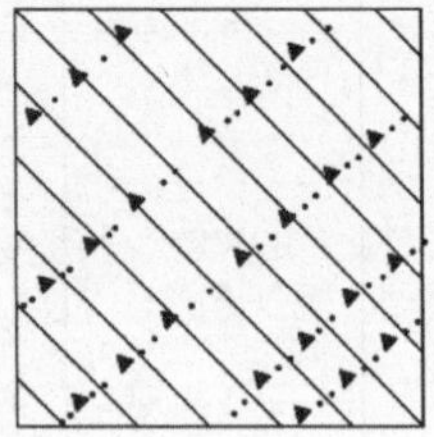
PLASTER EXISTING	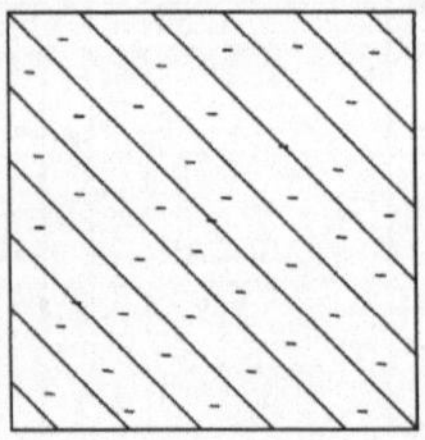	LIGHT WEIGHT CONCRETE	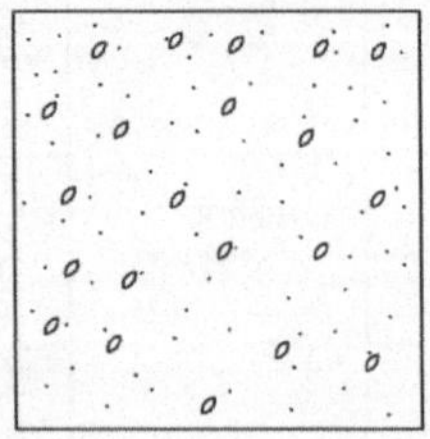
PLASTER ON STONE MASONARY	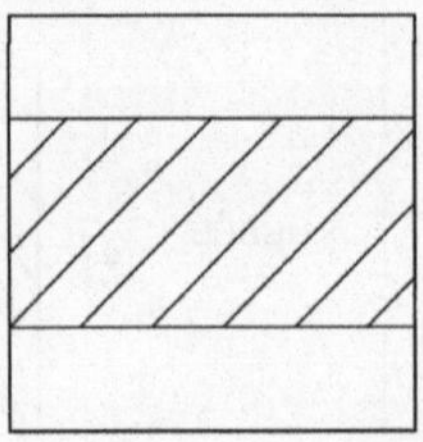	SAND	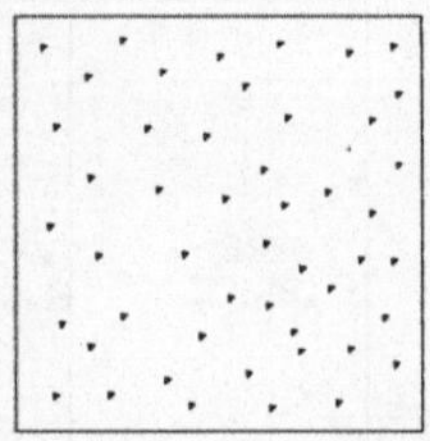
SOLID PLASTER PARTATION		MORRUM (OR) MOULDER	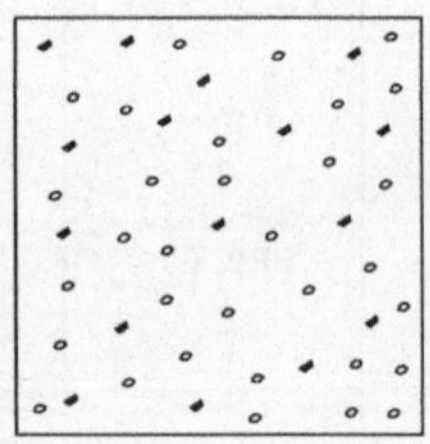
CONCRETE	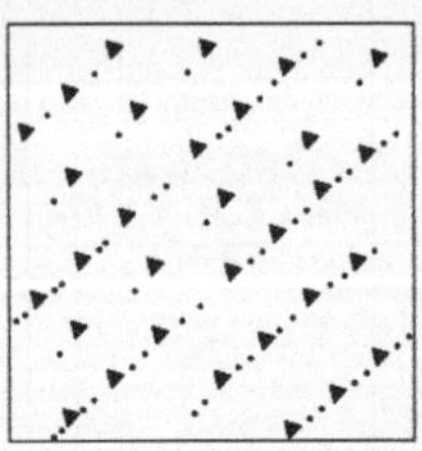	LIME TERRACE	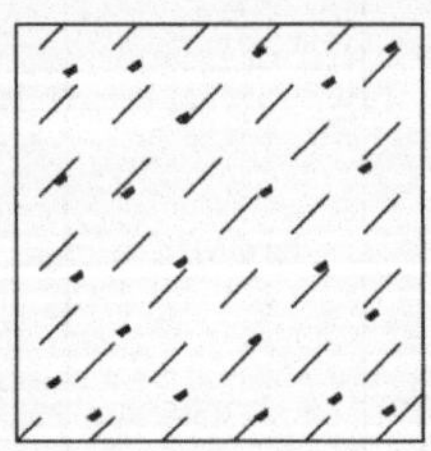

ELEMENT	SYMBOL	ELEMENT	SYMBOL
FLOOR FINISH	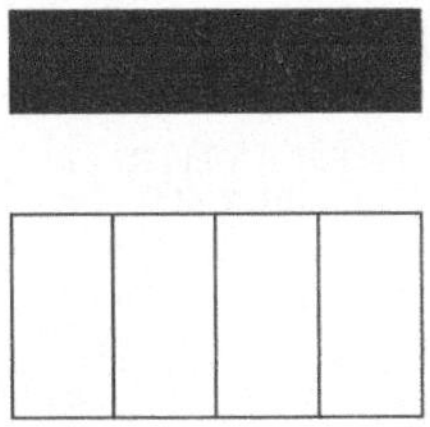	CLAY TILE (UN GLAZED)	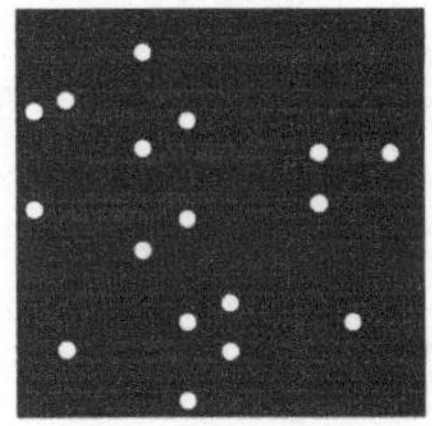
STEEL CASTIRON COPPER ALLOY	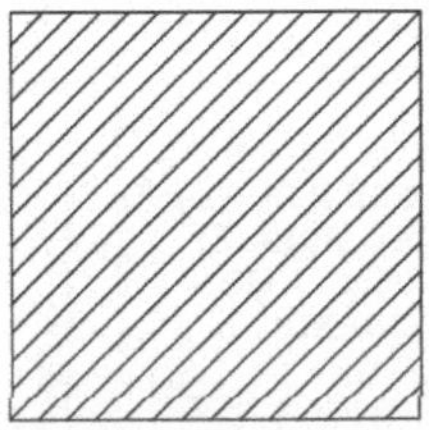	CERAMIC TILES	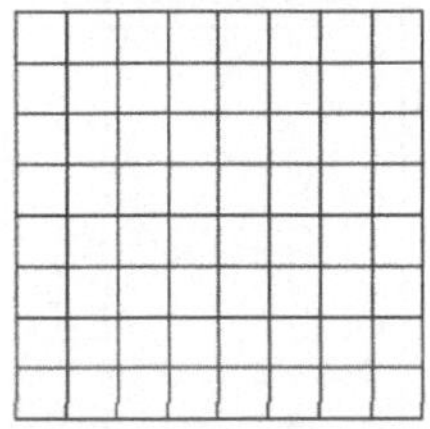
ALUMINIUM		EARTH	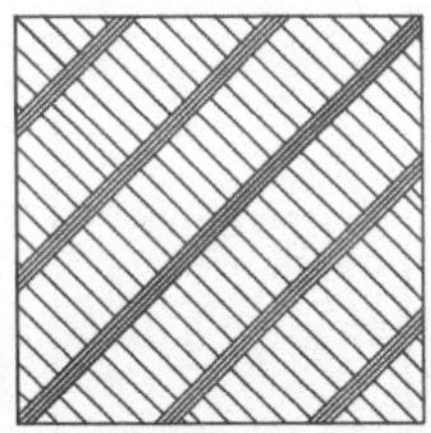
BRASS,BRONZE	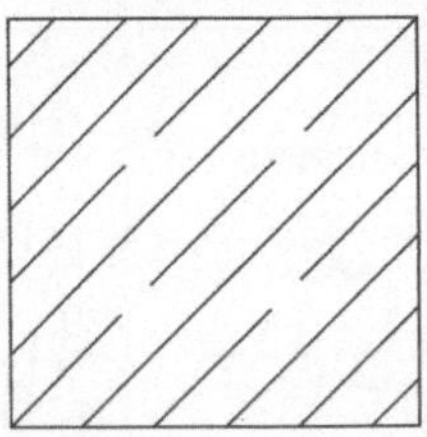	ROCK	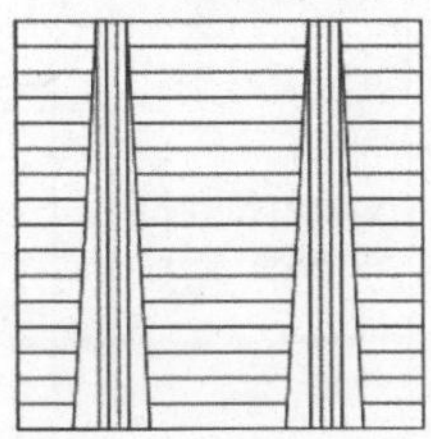
CLAY TILES (GLAZED)	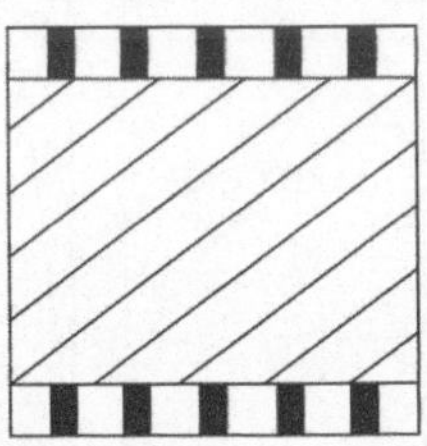	PLYWOOD	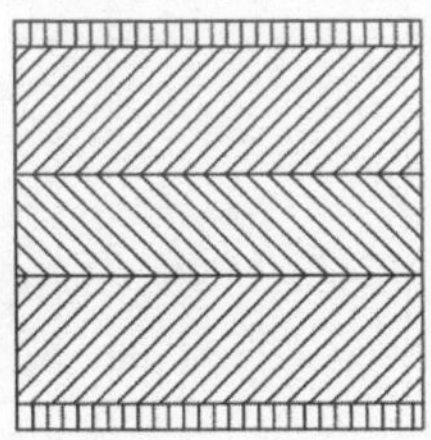

ELEMENT	SYMBOL	ELEMENT	SYMBOL
ROUGH WOOD	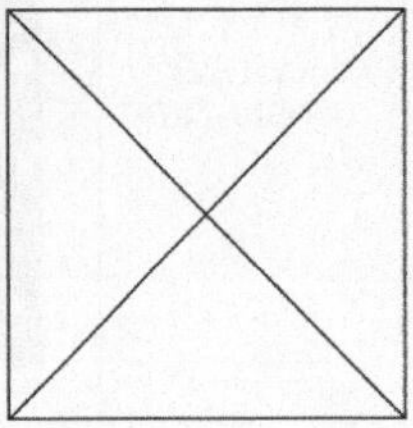	STEEL, COPPER ALUMINIUM	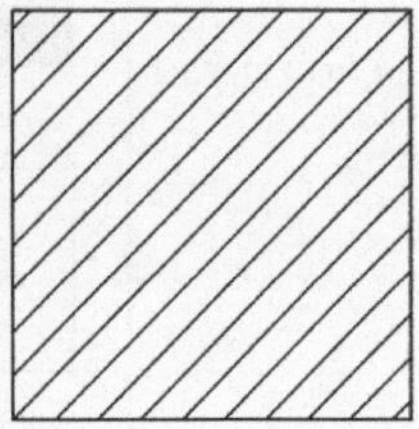
WOOD ACROSS GRAIN	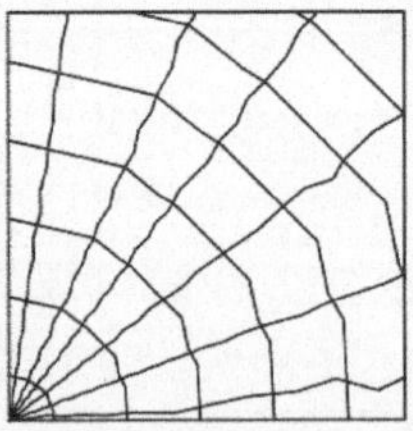	FIBER BUILDING BOARD AND INSULATION BOARD	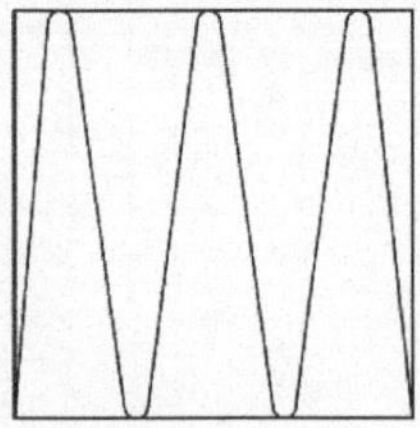
WOOD ACROSS GRAIN		CINDERS	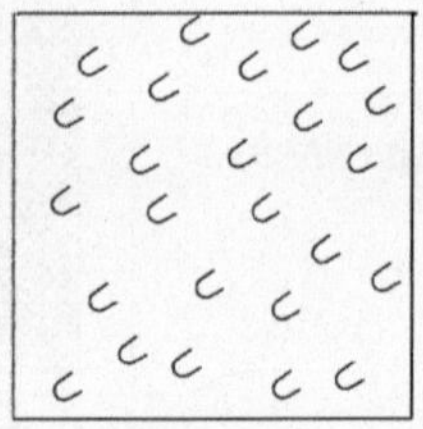
WOOD ALONG GRAINS	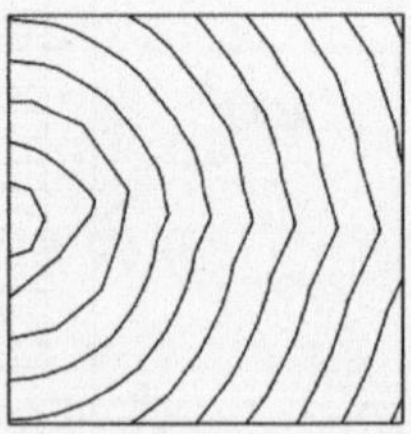	SHEET METAL	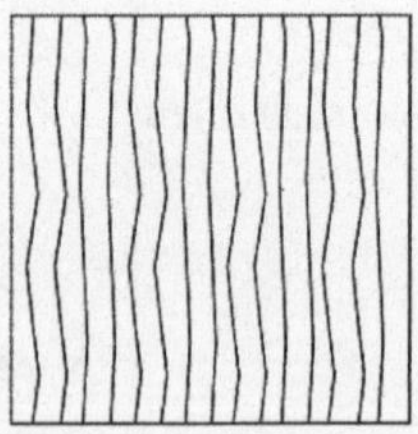
LEAD,ZINC	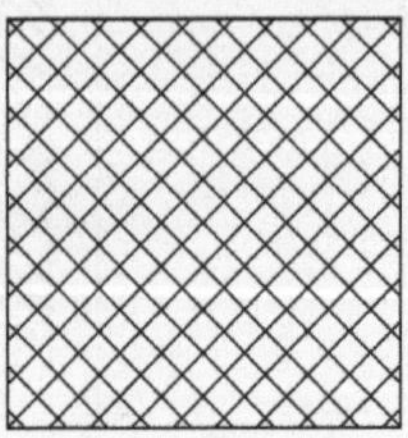	HARD CORE	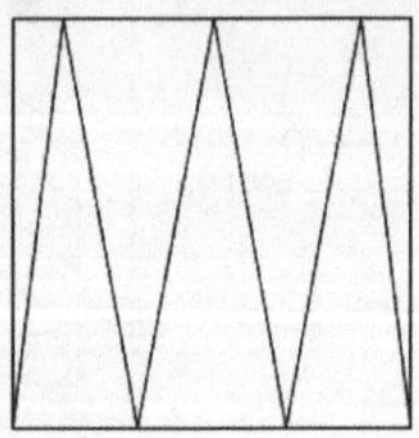

ELEMENT	SYMBOL	ELEMENT	SYMBOL
SOLID CORK AND PARTITION BLOCKS	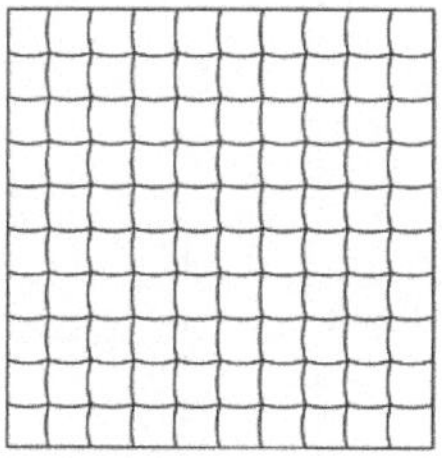	ASBESTOS, FIBER , CROCK, RUBBLE	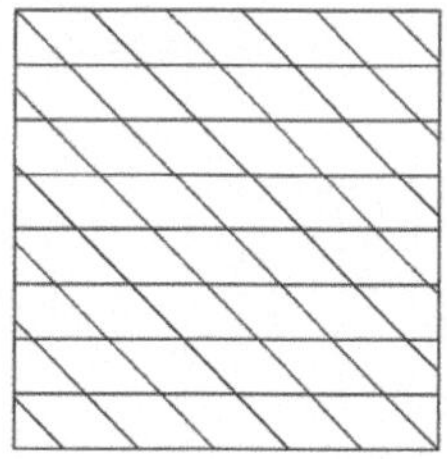
GLASS	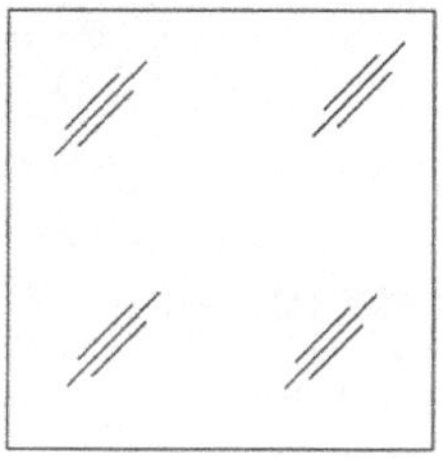	WOOD ALONG GRAINS	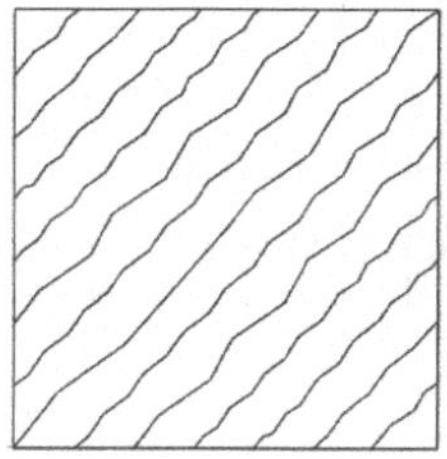

Fig. 2.1: Sign Conventions

2.2 ARCHITECTURAL

Architects need to be familiar with the various kinds of architectural signs and symbols that are frequently used in drawings. The general public who are involved in construction projects must be able to grasp all of the architect's produced drawings. Architectural symbols are needed for architectural drawings to be quickly and easily understood. As a result, the symbols and signs are created in such a way that they represent the finished work in terms of appearance and material. Architectural symbols are divided into six categories.

1) Door symbols 2) Window symbols 3) Wall symbols

4) Kitchen Symbols 5) Topography symbols 6) Bed room and Bath room symbols

2.2.1 Door Symbol

ELEMENT	SYMBOL	ELEMENT	SYMBOL
SINGLE LEAF SINGLE SWINGING DOOR	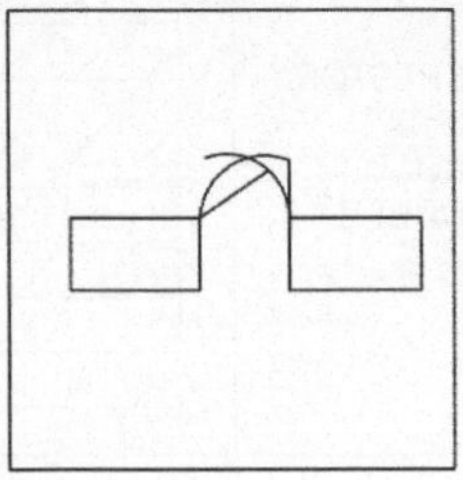	SLIDING DOOR	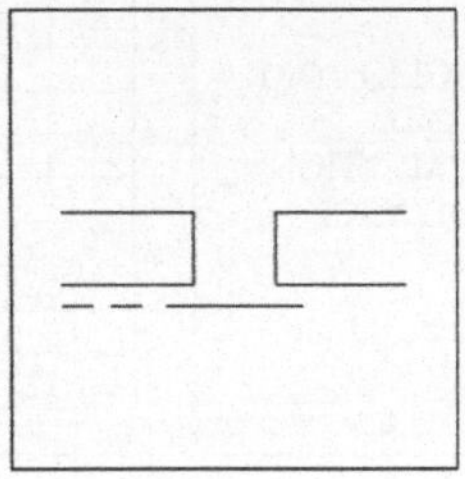
SINGLE LEAF DOUBLE SWINGING DOOR	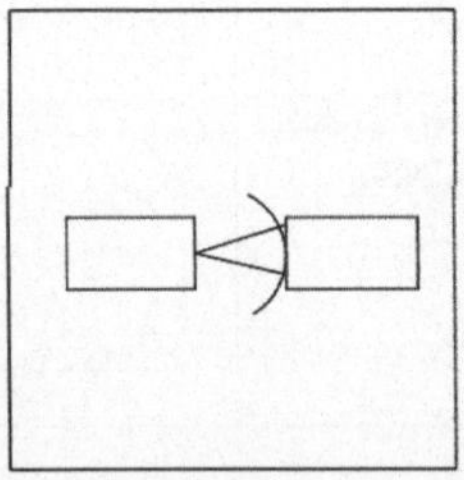	INTERNAL ROLLING SHUTTER	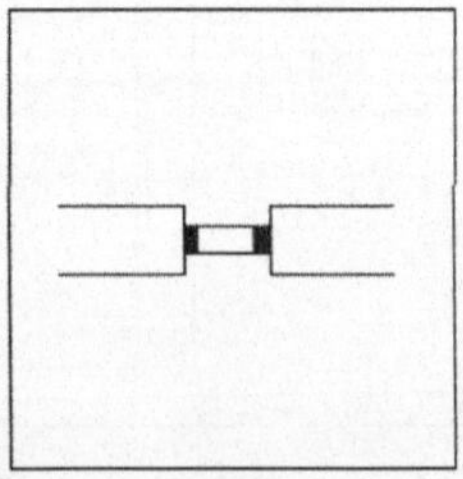
DOUBLE LEAF SINGLE SWINGING DOOR	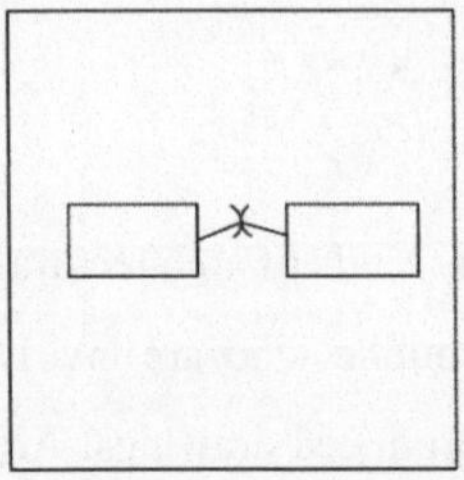	EXTERNAL ROLLING SHUTTER	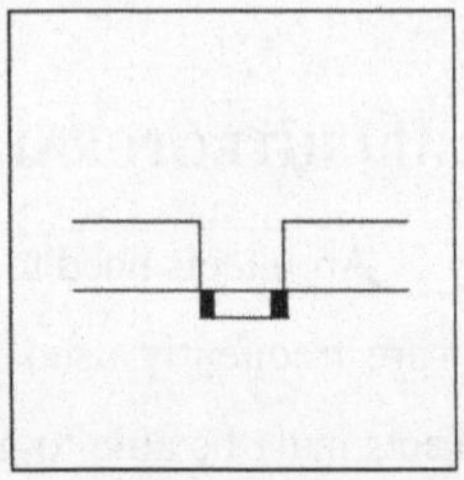
DOUBLE LEAF DOUBLE SWINGING DOOR	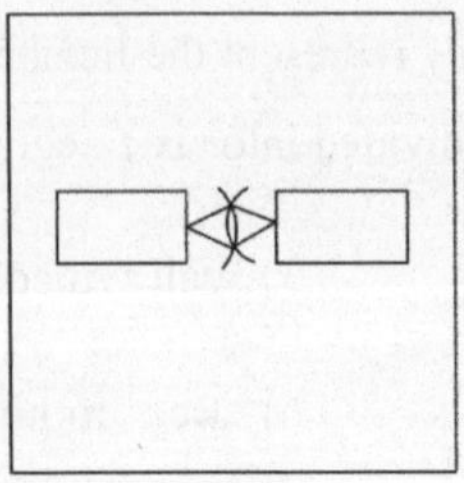	REVOLVING DOOR	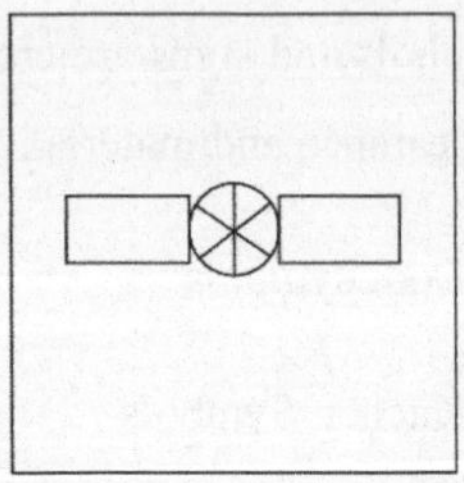

ELEMENT	SYMBOL	ELEMENT	SYMBOL
DOOR SWING	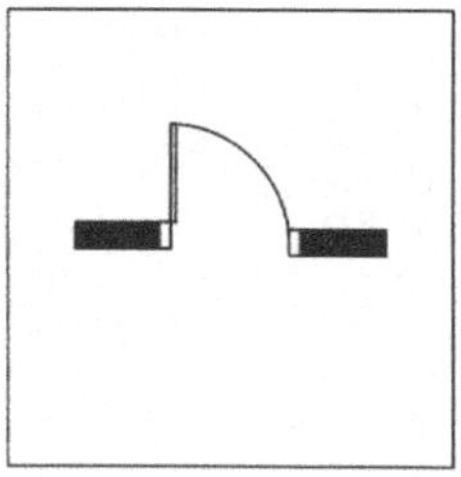	BYPASS SLIDING DOORS	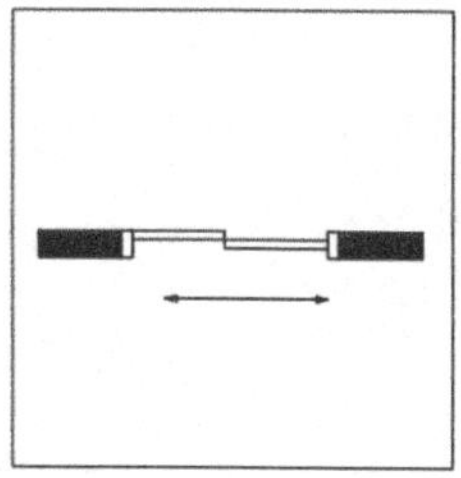
DOOR SWING - CLOSET DOOR	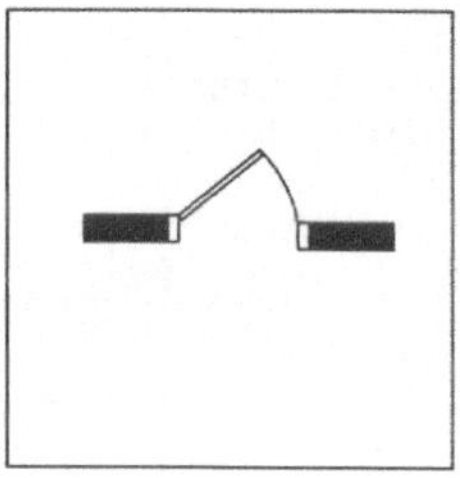	SURFACE SLIDING DOORS	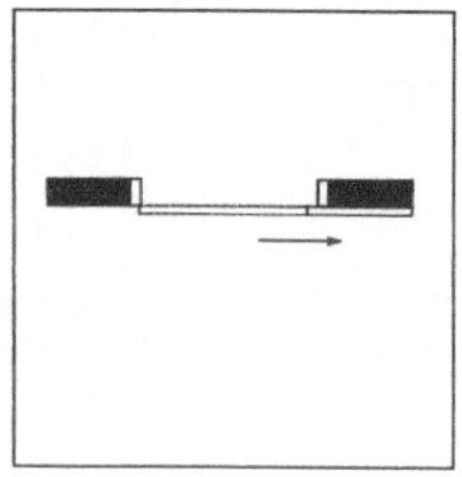
DOUBLE DOOR SWING	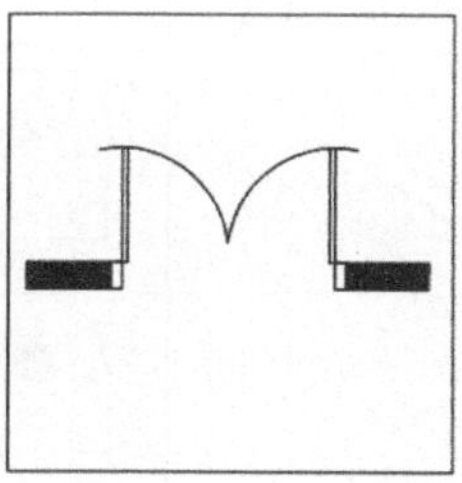	ARCHWAY-WALL OPENING	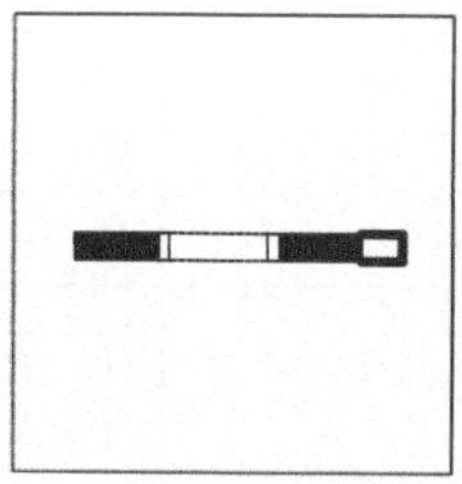
BIFOLD DOOR	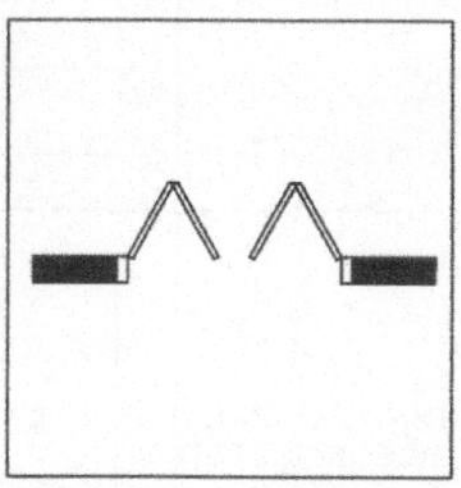	POCKET DOOR	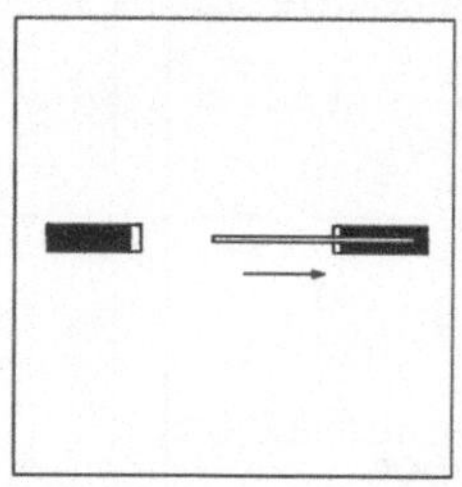

ELEMENT	SYMBOL	ELEMENT	SYMBOL
SINGLE DOOR, OPENING INWARDS	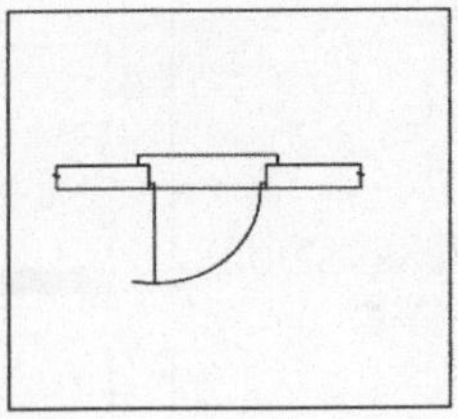	DOUBLE DOOR, OPENING INWARDS	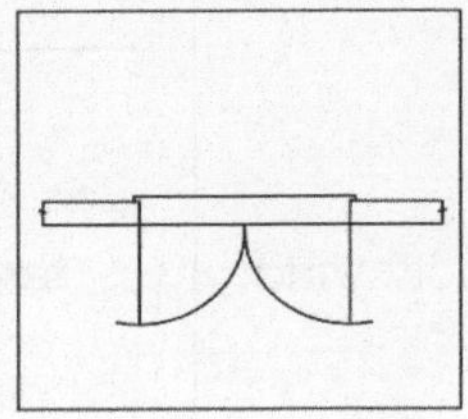
SINGLE DOOR, OPENING OUT	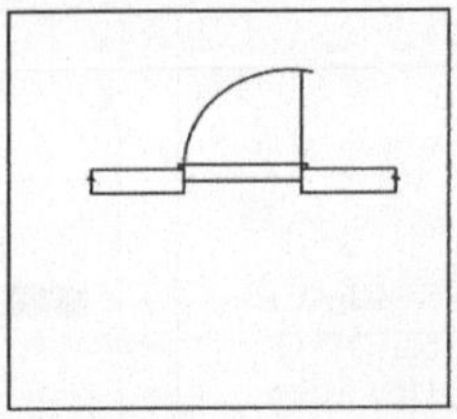	DOUBLE DOOR, OPENING OUT	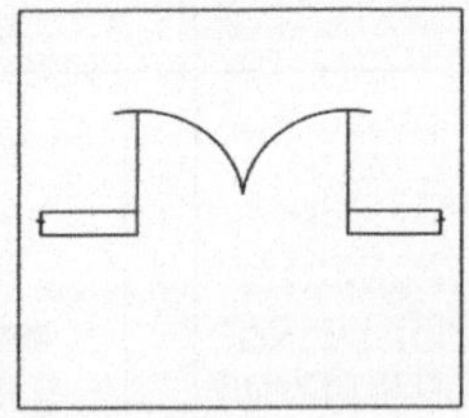
SINGLE DOOR, INTERIOR	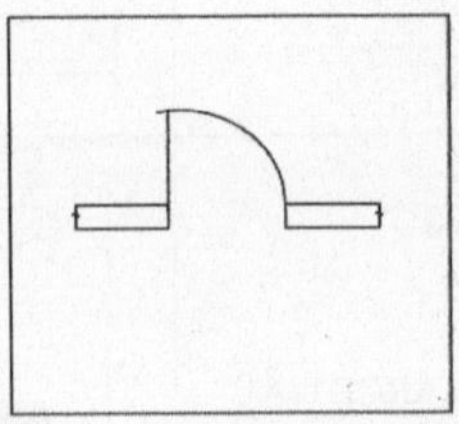	IN AND OUT DOORS	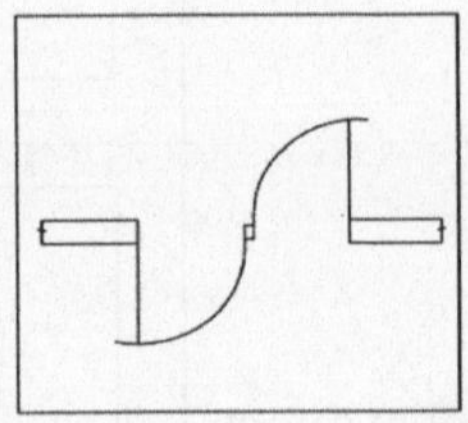
DOUBLE -ACTING SINGLE DOOR	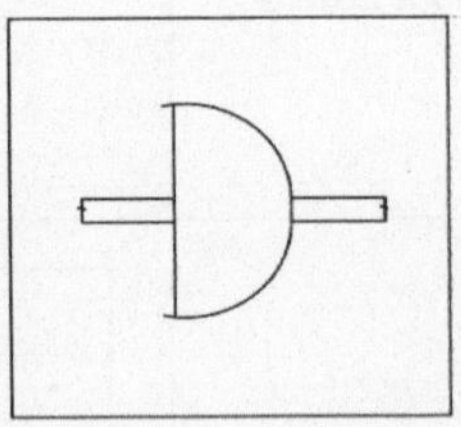	STEEL, COPPER ALUMINIUM	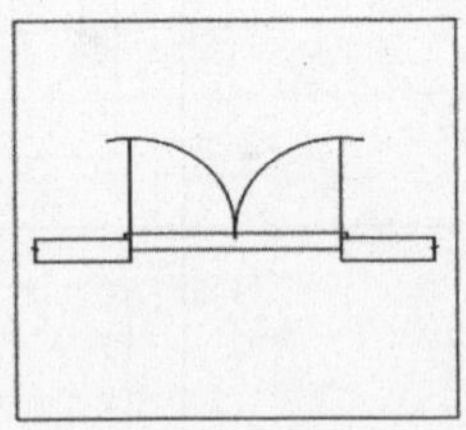
REFRIGERATOR DOOR	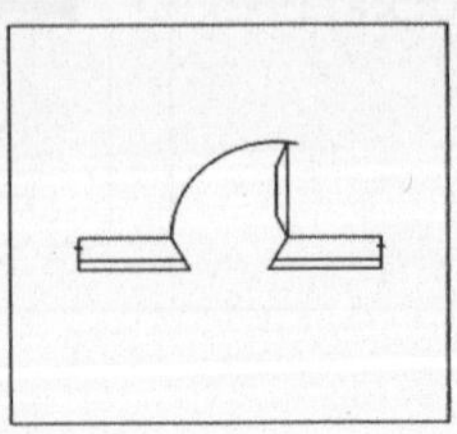	SLIDING DOOR-INTERCONNECTED	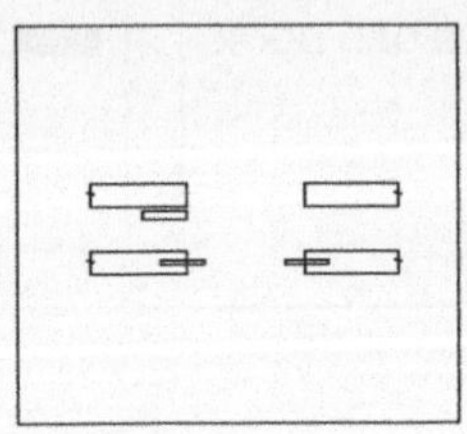

Fig. 2.2: Architectural Door Symbols

2.2.2 Window Symbol

ELEMENT	SYMBOL	ELEMENT	SYMBOL
HORIZONTAL CENTRAL HUNG WINDOW	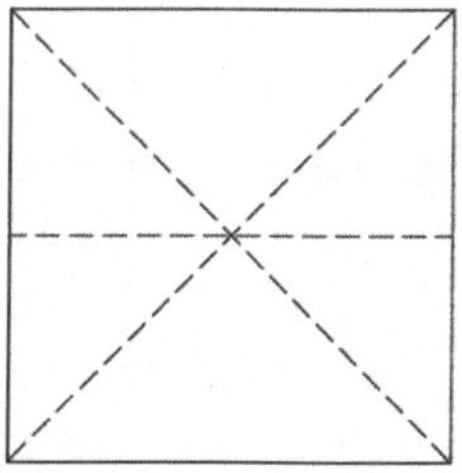	VERTICAL CENTRE HUNG WINDOW	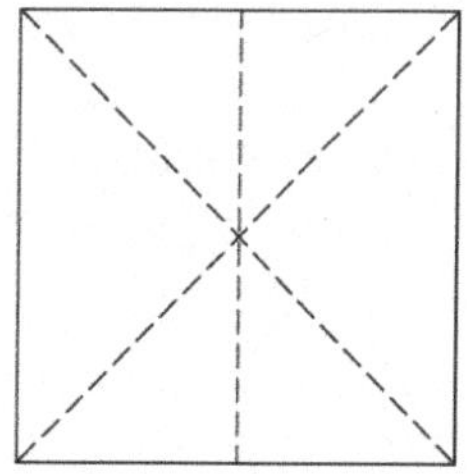
TOP HUNG WINDOW	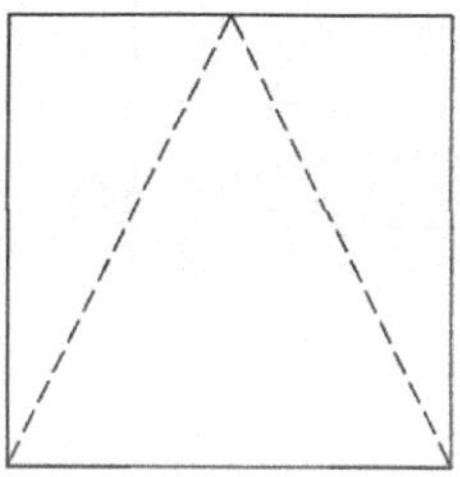	BOTTAM HUNG WINDOW	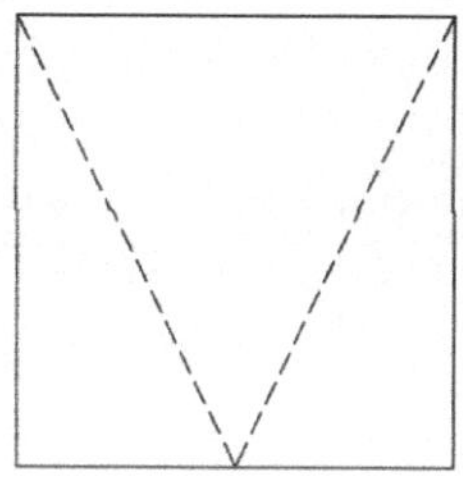
SIDE HUNG RIGHT HAND WINDOW	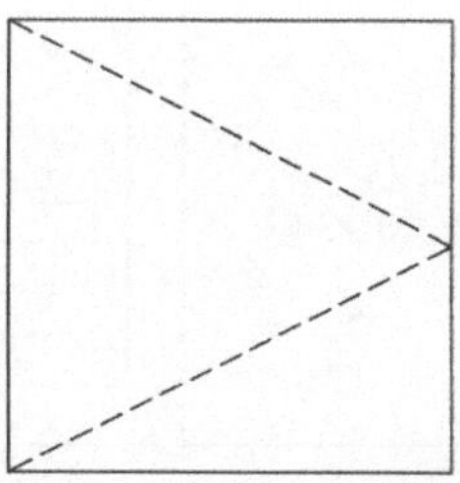	SIDE HUNG LEFT HAND WINDOW	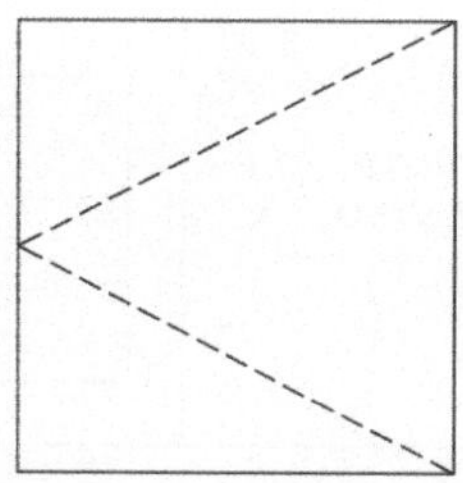

ELEMENT	SYMBOL	ELEMENT	SYMBOL
DOUBLE HUNG	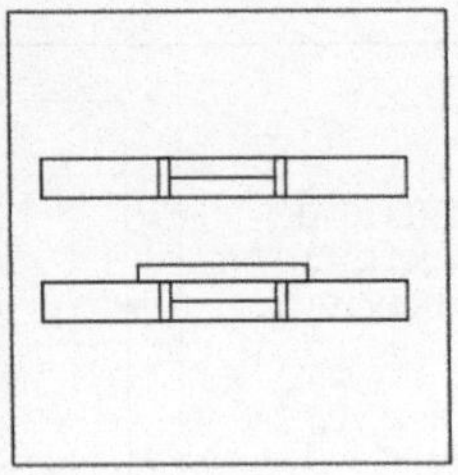	LEFT SASH OVER RIGHT	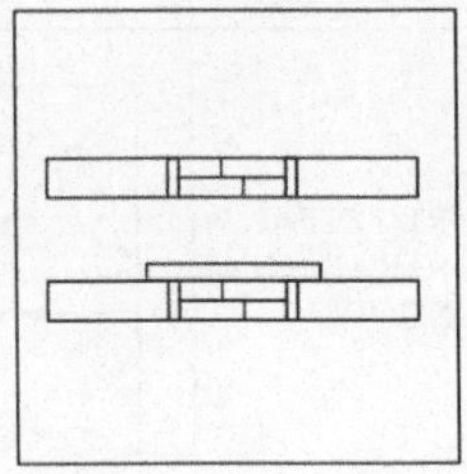
DOUBLE, OPENING OUT	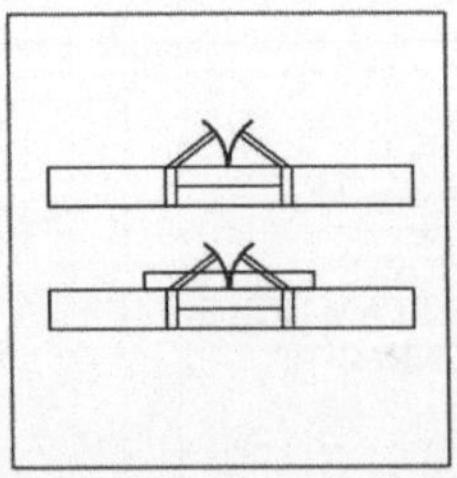	RIGHT SASH OVER LEFT	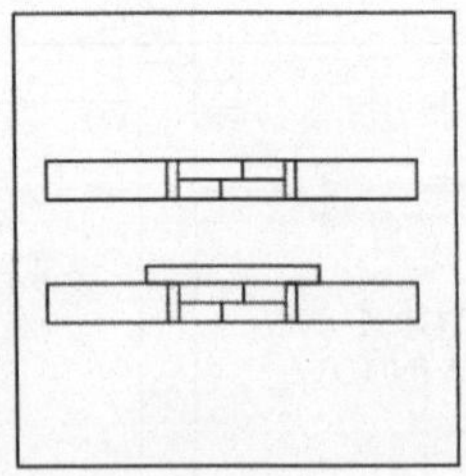
PIVOTED AND VENTED	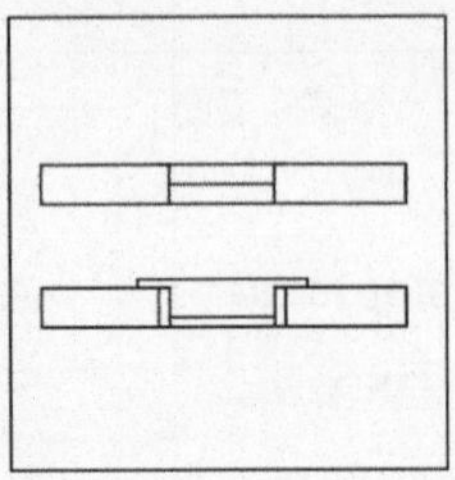	SINGLE, OPENING IN	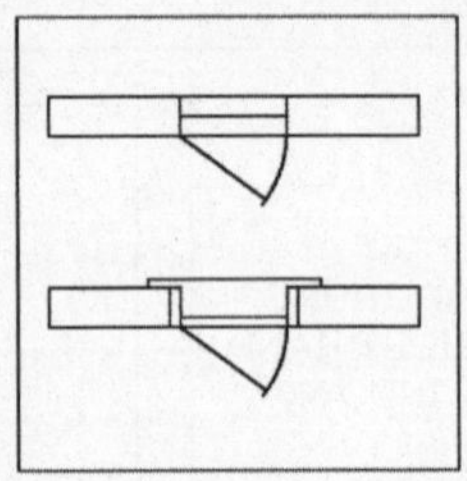

Fig. 2.3: Architectural Window Symbols

2.2.3 Wall Symbols

ELEMENT	SYMBOL	ELEMENT	SYMBOL
PROPERTY LINE		WALL	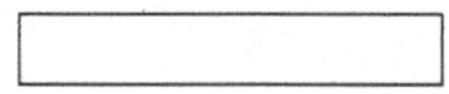
CENTER LINE		STONE WALL	
BUILDING		FENCE	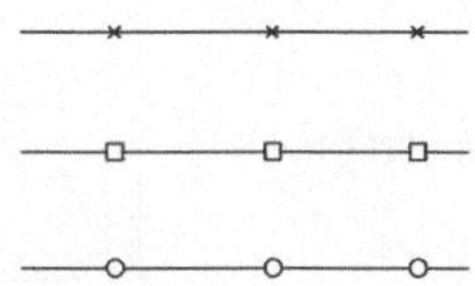

Fig.2.4: Architectural Wall Symbols

2.2.4 Kitchen Symbols

ELEMENT	SYMBOL	ELEMENT	SYMBOL
PLAIN KITCHEN SINK	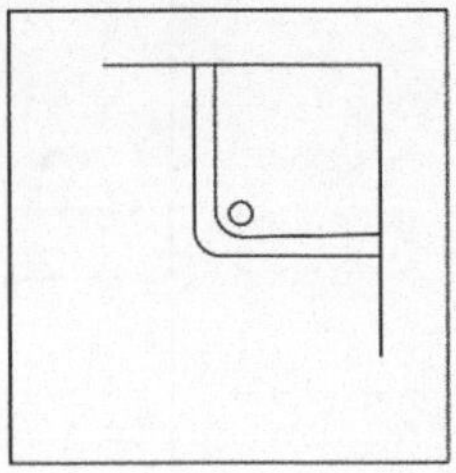	COOKER	C
KITCHEN SINK WITH DRAIN	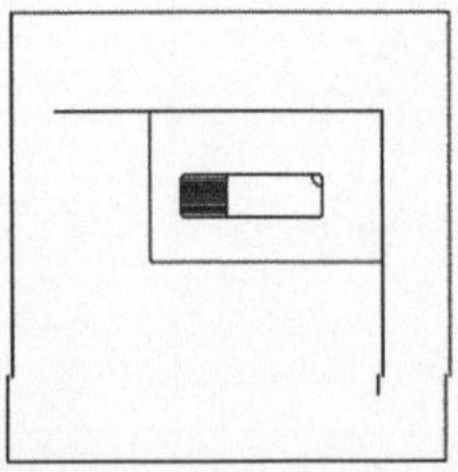	REFRIGERATOR	
WASH DRAIN	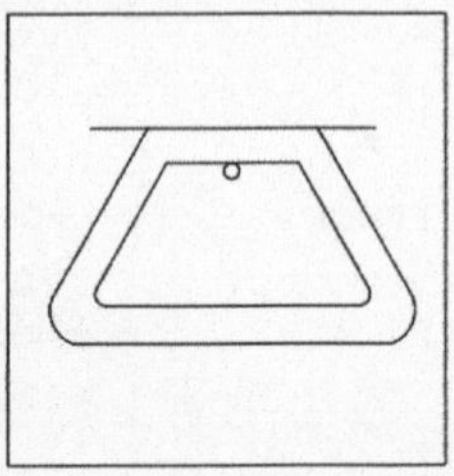	DRINKING WATER FOUNTAIN	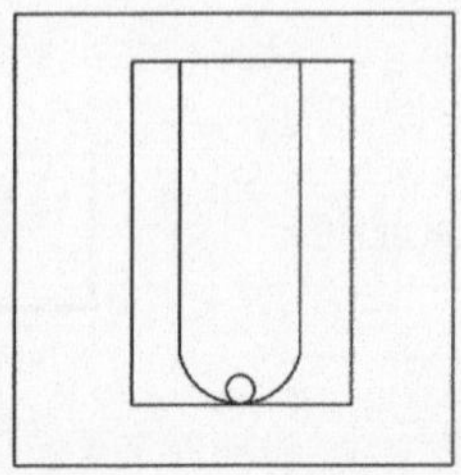
COOKING PLATFORM	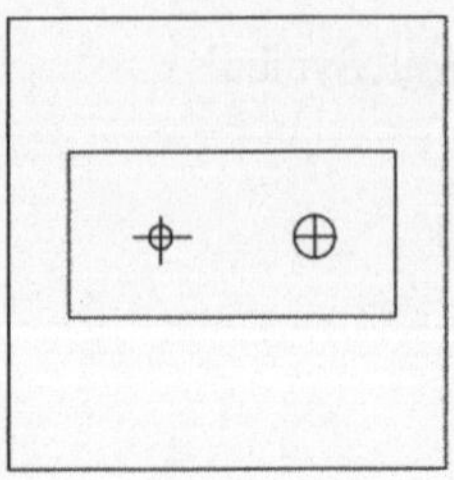	R.W.O. RAIN WATER OUTLET	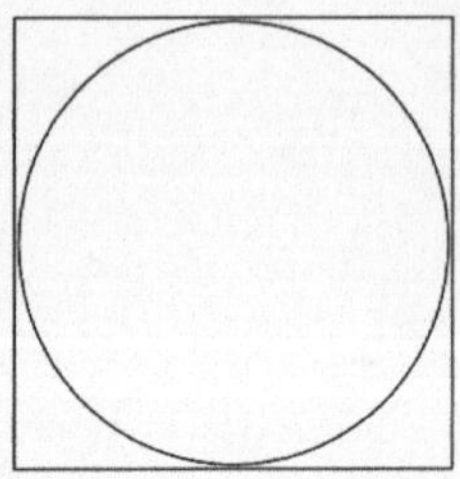

Fig.2.5: Architectural Kitchen Symbols

2.2.5 Bed Room Symbols

ELEMENT	SYMBOL		ELEMENT	SYMBOL
BED	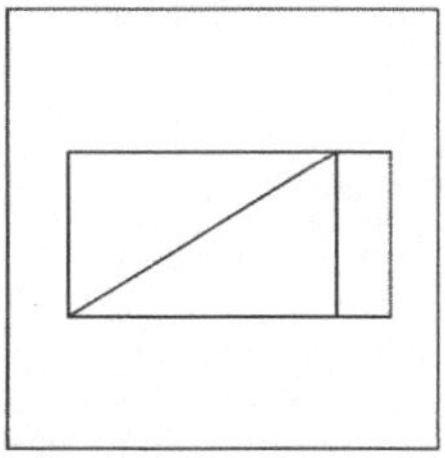		ALMIRAH	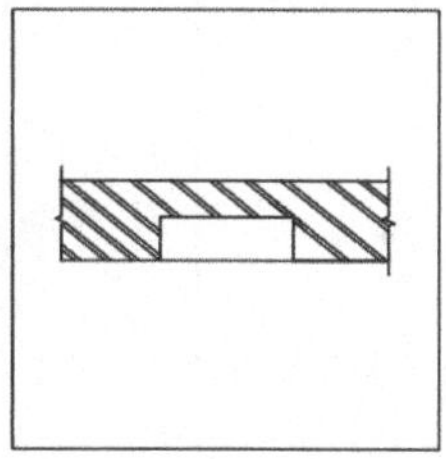
CUP BOARD	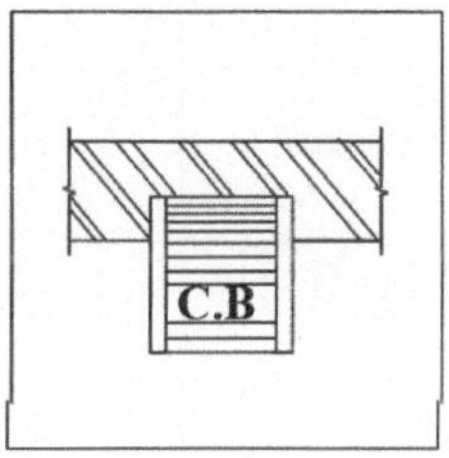		CIRCULAR BED	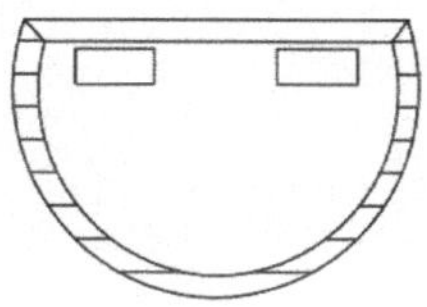

Fig. 2.6: Architectural Bed room and Bath room Symbols

2.2.6 Topography Symbols

ELEMENT	SYMBOL	ELEMENT	SYMBOL
ROCK	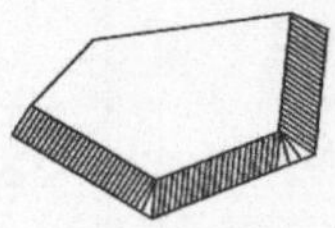	SANITARY SEWERS	
SWAMP	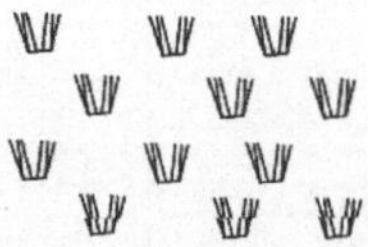	GAS LINE	
TANK (LABLE TYPE) SEPTIC		WATER LINE	
SINGLE, OPENING IN	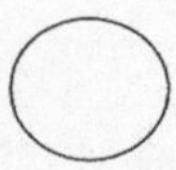	UNPAVED ROAD	
SURVEYED CONTOURS	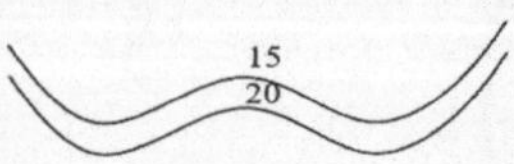	PAVED ROAD	

ELEMENT	SYMBOL	ELEMENT	SYMBOL
RAIL ROAD TRACKS		BUSHES	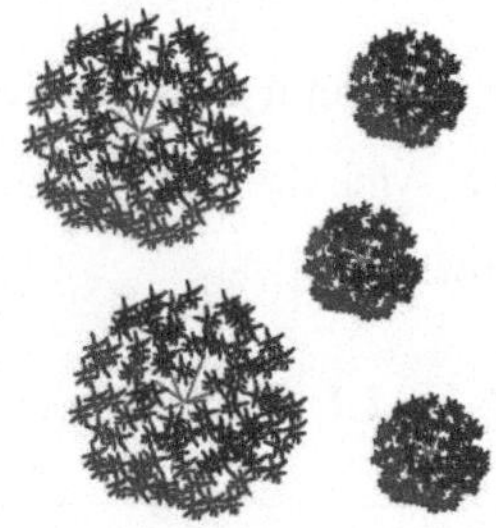
MARSH	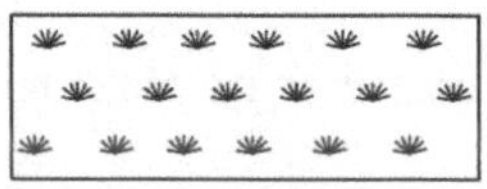	TREES	
GROUND COVER	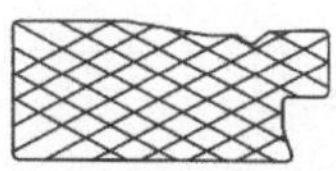	GRASS	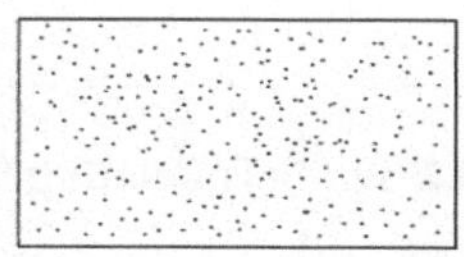
NORTH ARROW		TALL GRASS	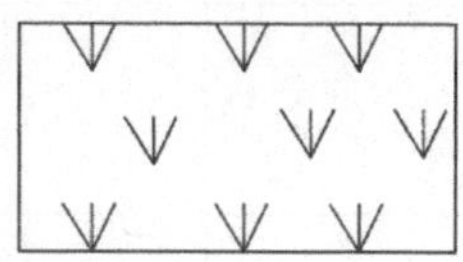

Fig. 2.7: Architectural Topographical Symbols

2.3 STRUCTURAL

2.3.1 Symbols Relating to Cross-sectional Shape and Size of Reinforcement

In Structural drawings, reinforcement details will often be accompanied by notes that explain the symbols used, including the materials, diameter, & specific configurations about the project.

Table 2.1: Symbols relating to reinforcement

SYMBOL	DESCRIPTION
Ø	Plain round bar or diameter of plain round bar
□	Plain square bar or side of plain square bar
#	Deformed bar (including square twisted bar) or nominal size (equivalent diameter or side) of the deformed bar

2.3.2 Symbols Relating to Position and Direction

Table 2.2: Symbols Relating to Position and Direction

SYMBOL	DESCRIPTION
EW	Each way
@	Spacing centre-to-centre
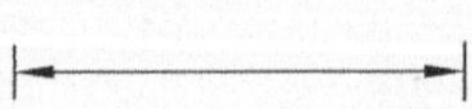	Limit of area covered by bars
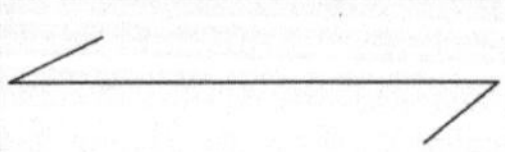	Direction in which bars extend

2.3.3 Symbols Relating to Shape of the bar along its length

It is the guide to the various reinforcement detailing symbols:

Table 2.3: Symbols relating to shape of reinforcement

SYMBOL	DESCRIPTION
Alt	Alternate bar
Bt	Bent bar
B	Bottom bar
min	Minimum
max	Maximum
St	Straight bar
Stp	Stirrup bar
Sp	Spiral
Ct	Column tie
T	Top bar

2.3.4 Symbols Relating to Various Structural Members

Table 2.4: Symbols relating to structural members

SYMBOL	DESCRIPTION
Bm or B	Beams
Col	Columns
Fg	Footings
GR	Girders
JT	Joints
LL	Lintels
LB	Lintel beams
Sb or S	Slab
WL	Longitudinal wall
Wx	Cross wall
c	Centre line

2.3.5 Symbols Relating to Structural Drawings for Enhancing the clarity

Each symbol typically has an accompanying note or legend that provides details about the material, type of reinforcement which are represented by various lines.

Table 2.5: Symbols relating to structural drawings

SYMBOL	DESIGNATION/DESCRIPTION
	Concrete line (thin)
	Unexposed concrete or masonary wall line (thin)
	Reinforcement (thick)
	Reinforcement in a diffrent layer (thick)
	Section of reinforcing bar
	Center line
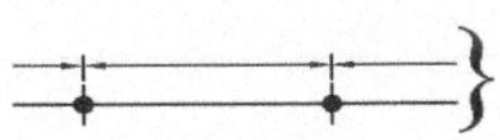	Dimension line
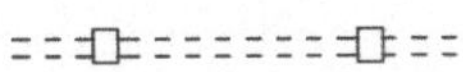	Concrete beam framing into column which extends through floor
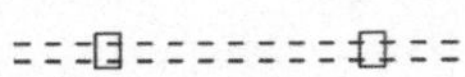	Concrete beam framing into column which stops at floor

SYMBOL	DESIGNATION/DESCRIPTION
	Bar shown bent at right angle to the paper
	Bar with hooks
	Bar with 90^0 bends
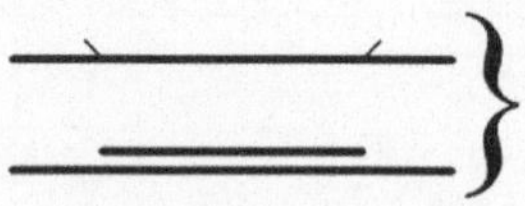	Bar shown separated on the drawing
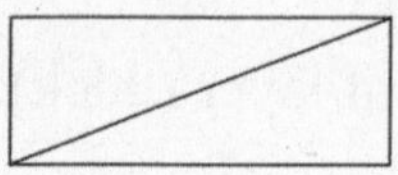	One sheet of welded fabric on plan
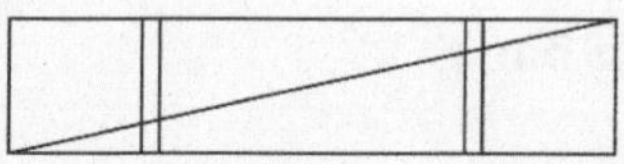	Identical sheets of welded fabric in a row
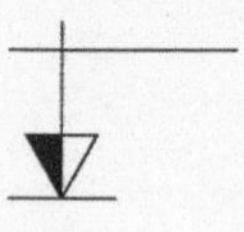	Level mark in elevation
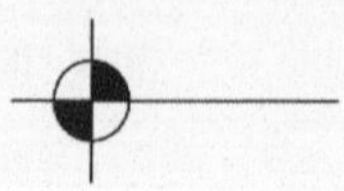	Level mark in plan

2.4 ELECTRICAL SYMBOLS

Electrical symbols are pictograms used to represent various electrical devices in a plan of an electrical circuit.

Table 2.6: Electrical symbols

ELEMENT	SYMBOL	ELEMENT	SYMBOL
FIRE EXTINGUISHER	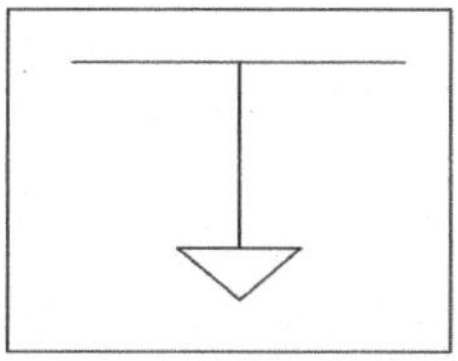	CEILING FAN	
SAFETY VALVE	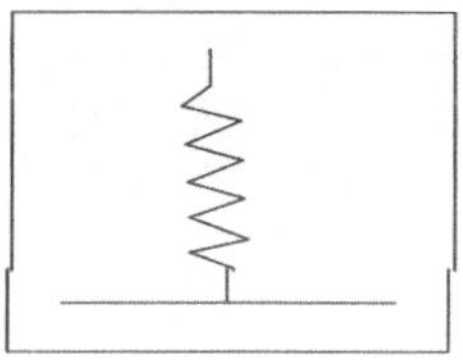	WATER METER	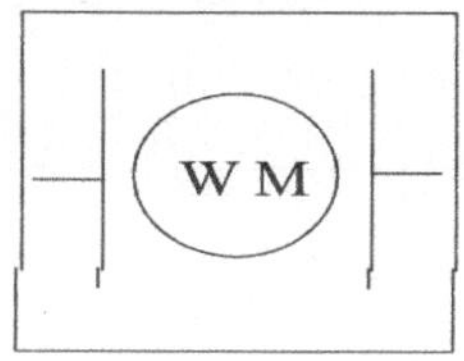
PUMP	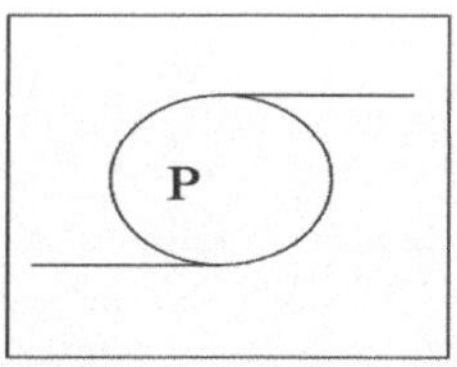	FIRE HYDRANT	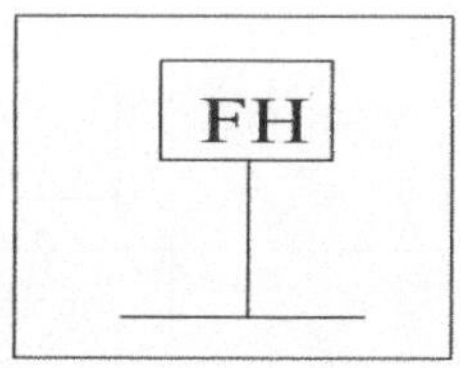
EARTH POINT	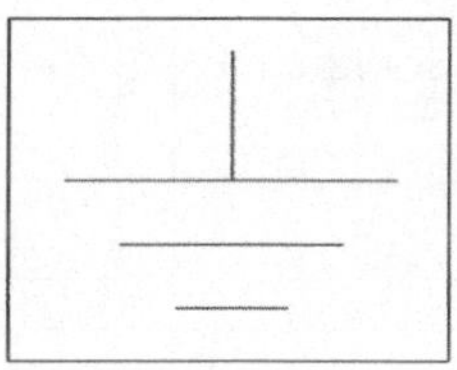	SINGLE POLE SWITCH 230V,10A	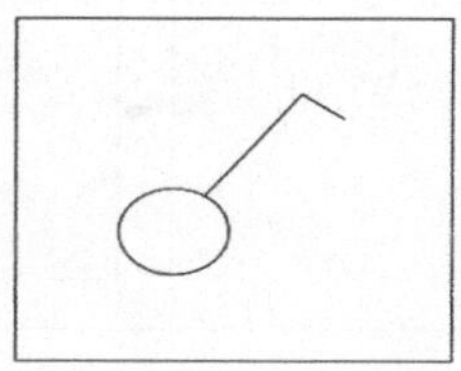
EXHAUST FAN	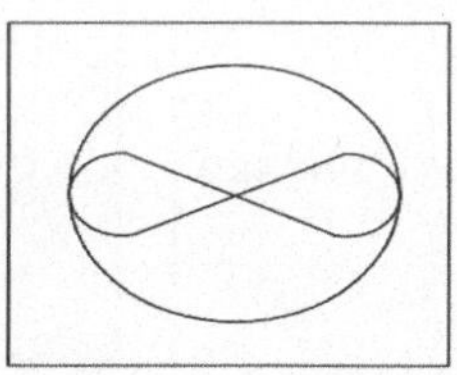	BUTTON DRIVER 230V,10A	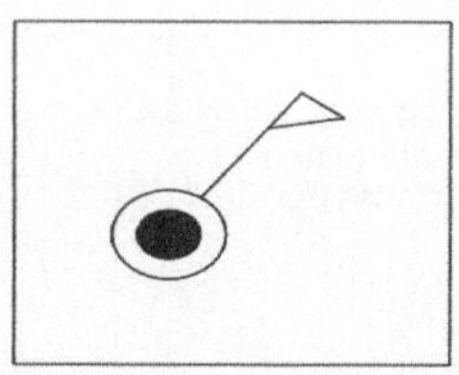

ELEMENT	SYMBOL	ELEMENT	SYMBOL
DOUBLE POINT SWITCH 230V,10A	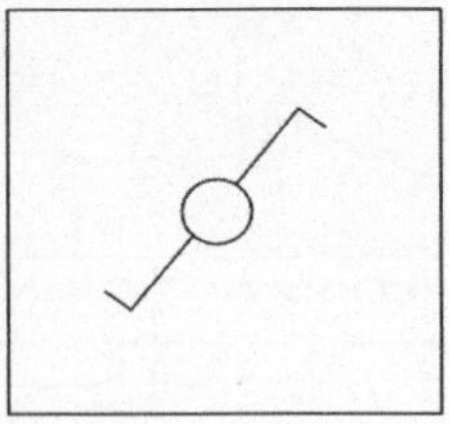	INFRARED ROOM SENSOR 10A,230V	
CROSS SWITCH 230V,10A	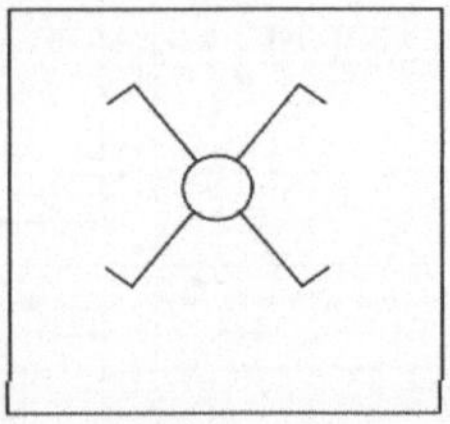	SOCKET 230V,16A	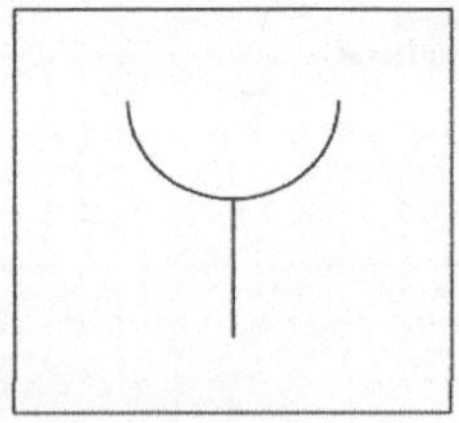
SERIAL SWITCH 230V,10A	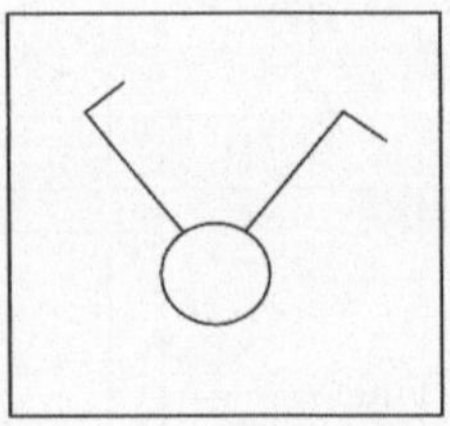	SOCKET IP65 230V,16A	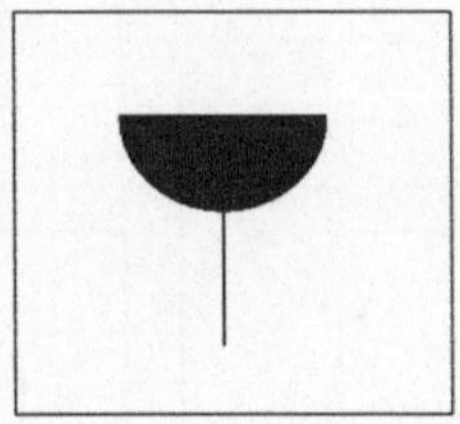
SERVICE SWITCH 400V,10A	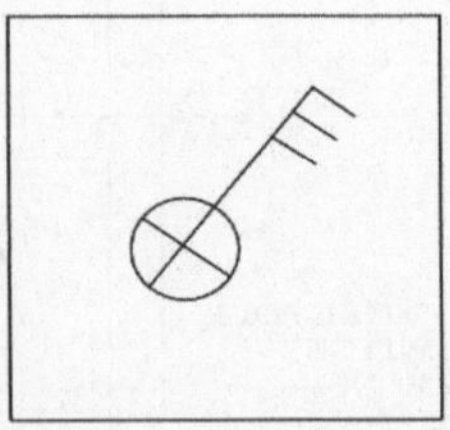	DOUBLE SOCKET 230V,16A	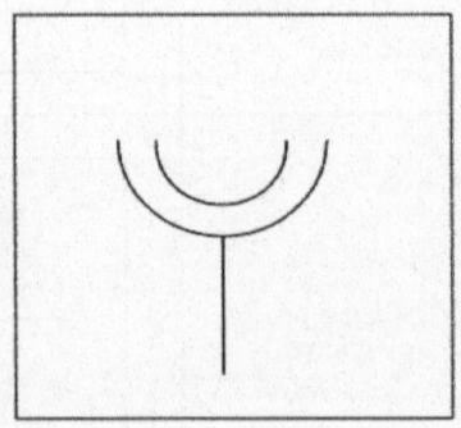
BLIND CONTROL 230V,10A	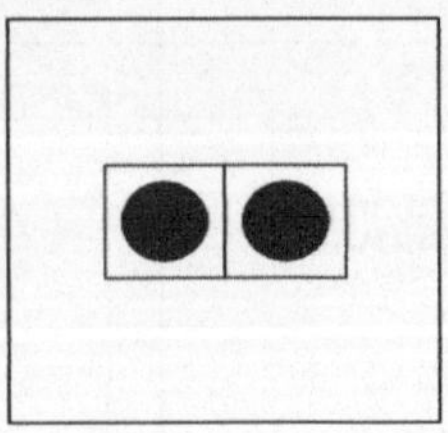	SWITCHBOARD	

ELEMENT	SYMBOL	ELEMENT	SYMBOL
SINGLE SWITCHED WALL SOCKET	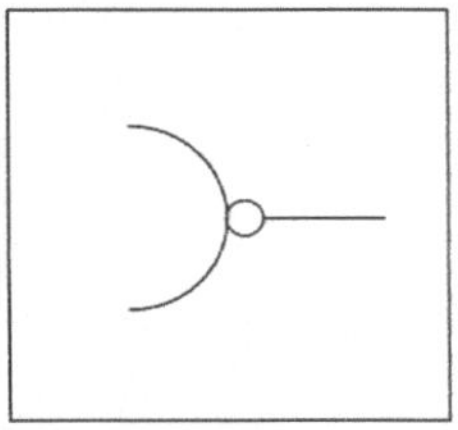	FUSE SPUR FOR TOWER RAIL	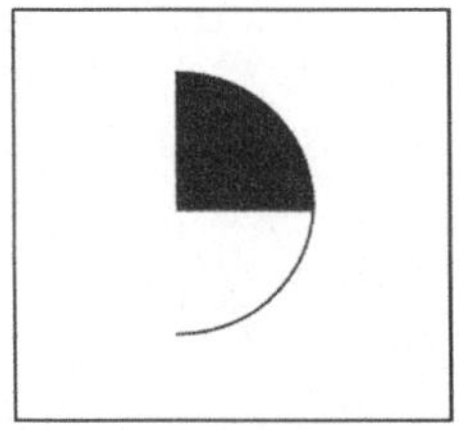
DOUBLE SWITCHED WALL SOCKET	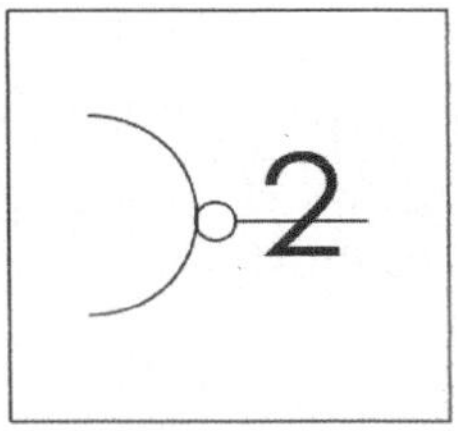	SINGLE SWITCHED FUSE SPUR AT HIGH LEVEL FOR EXTRACT FAN	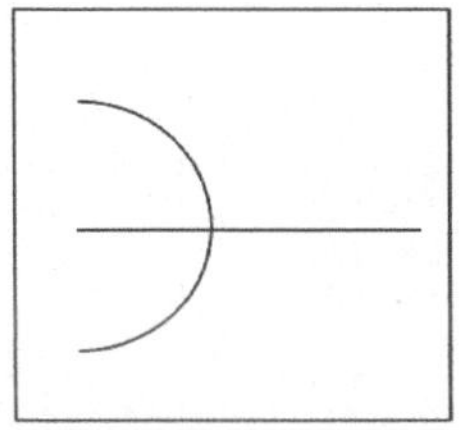
SINGLE SWITCHED WALL SOCKET 1200MM ABOVE FLOOR LEVEL	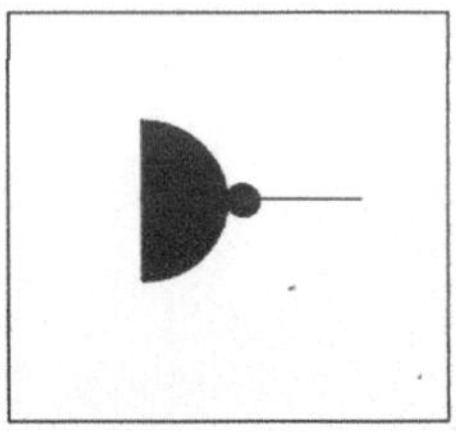	DOUBLE SWITCHED FLOOR SOCKETS	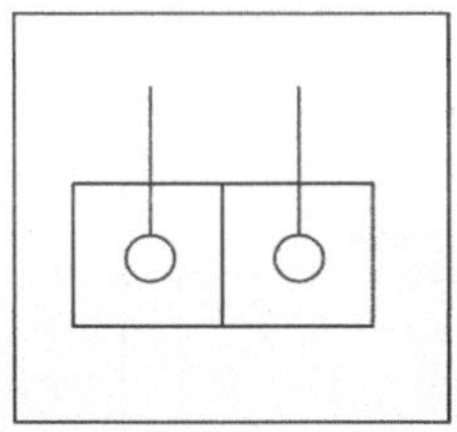
DOUBLE SWITCHED WALL SOCKET 1200MM ABOVE FLOOR LEVEL	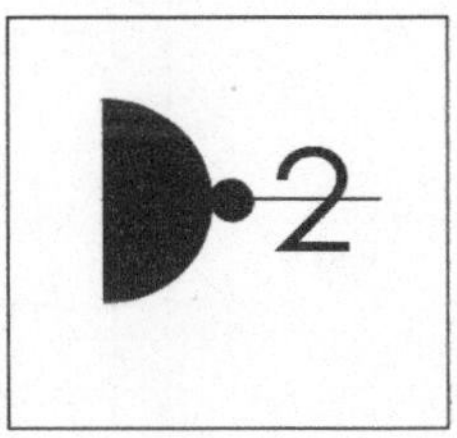	1 GANG LIGHT SWITCH	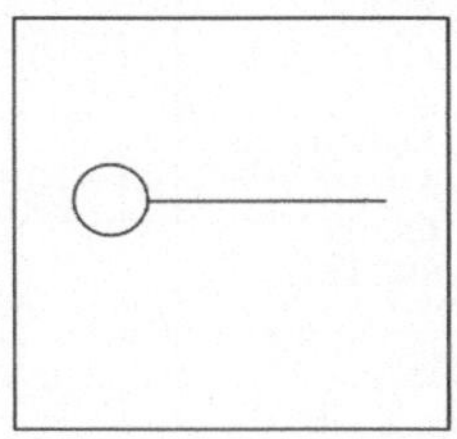
SINGLE SWITCHED FUSE SPUR BELOW WORK TOP LEVEL	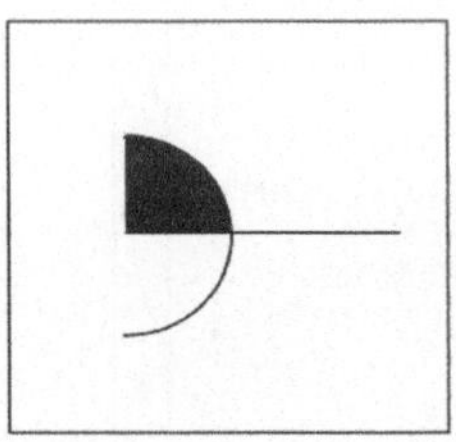	2 GANG LIGHT SWITCH	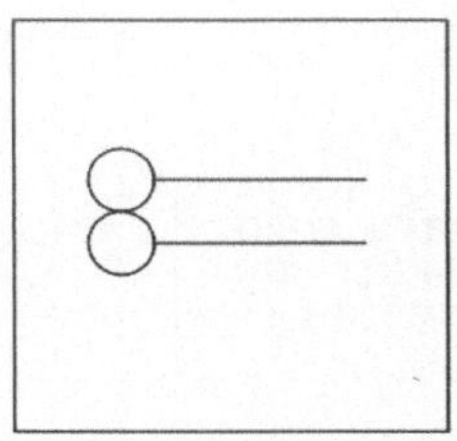

ELEMENT	SYMBOL	ELEMENT	SYMBOL
1 GANG 2 WAY LIGHT SWITCH		EXTERNAL LIGHT	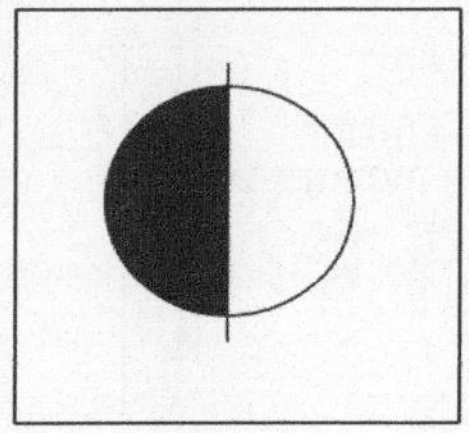
2 GANG 2 WAY LIGHT SWITCH	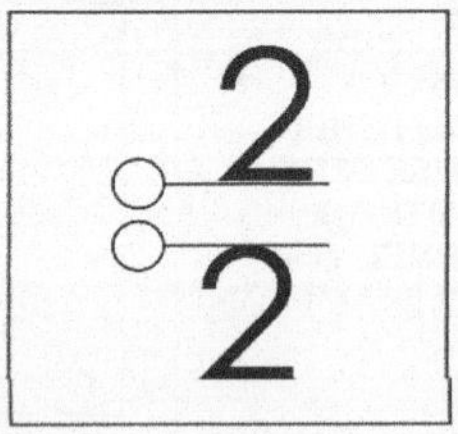	EXTERNAL LIGHT BULKHEAD TYPE	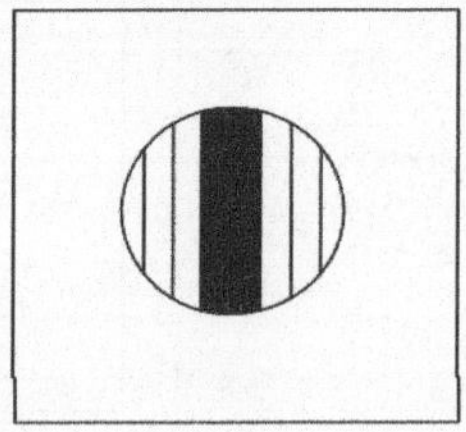
1 GANG 3 WAY LIGHT SWITCH	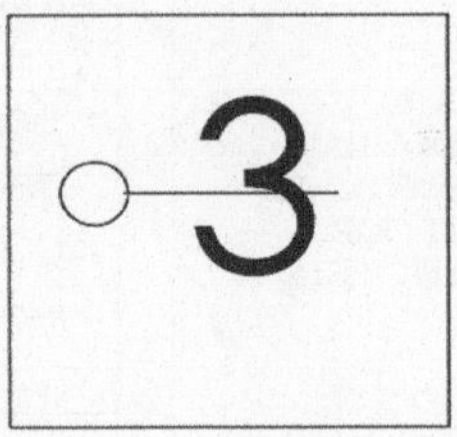	RECESSED DOWN LIGHTER	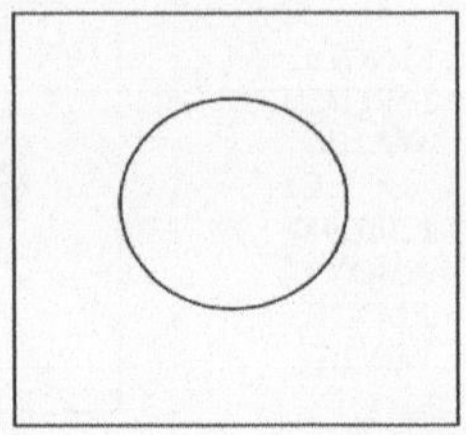
2 GANG 3 WAY LIGHT SWITCH	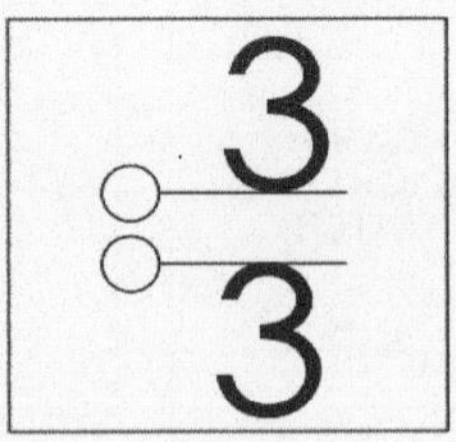	WALL LIGHT	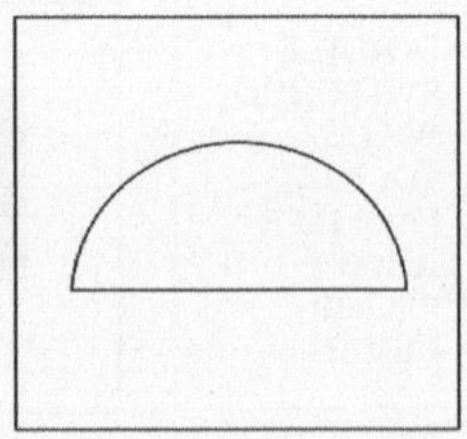
1 GANG PULL CORD LIGHT SWITCH	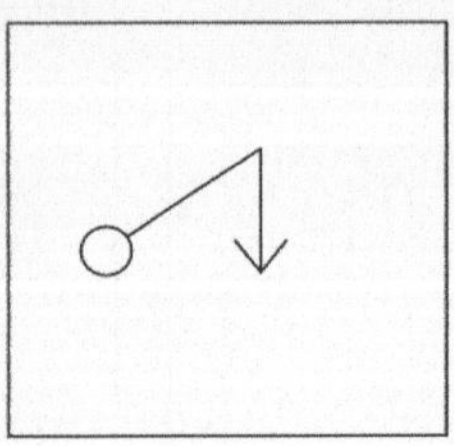	LIGHT POINT PENDANT HOLDER (4 INCH)	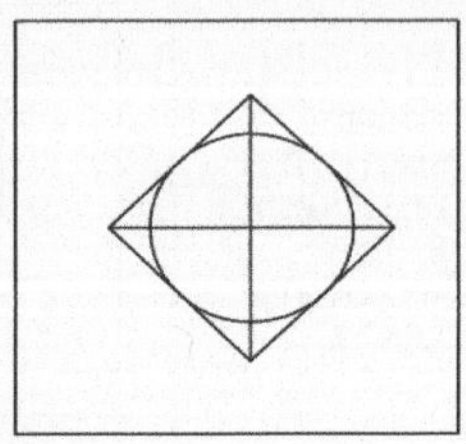

ELEMENT	SYMBOL	ELEMENT	SYMBOL
LIGHT POINT PENDANT HOLDER (6 INCH)	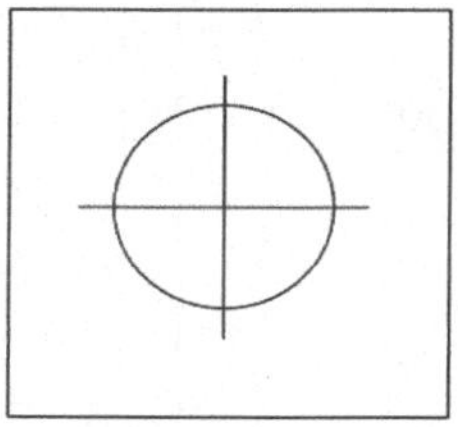	MECHANICAL EXTRACT FAN	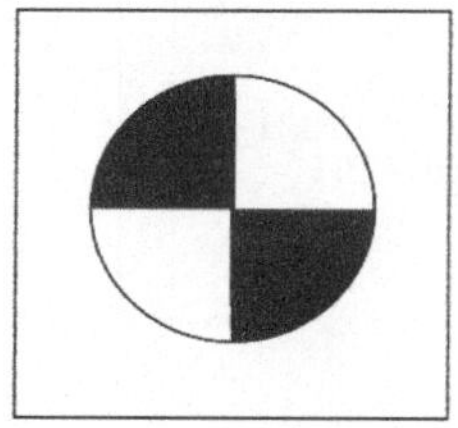
LIGHT POINT BATTEN HOLDER	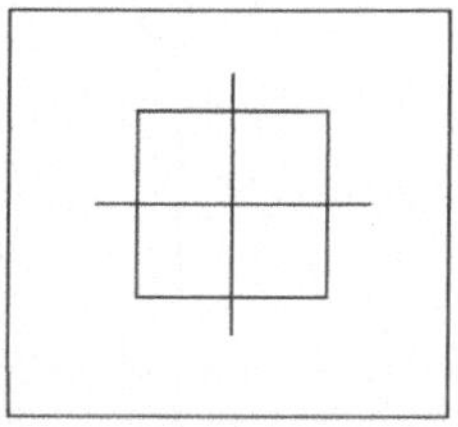	DOOR BELL PUSH	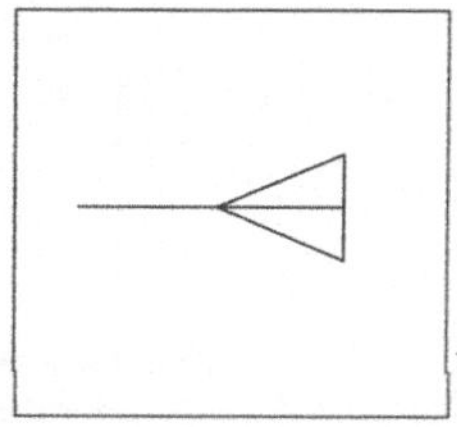
FLUORESCENT STRIP LIGHT	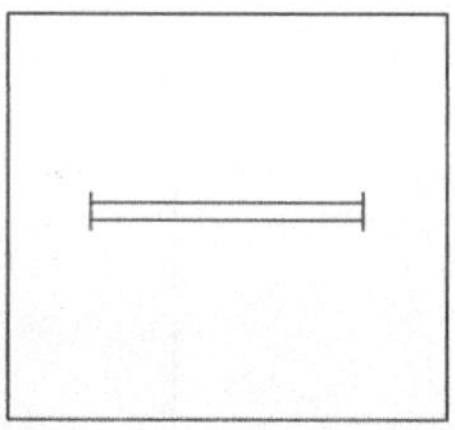	DOOR BELL CRIME	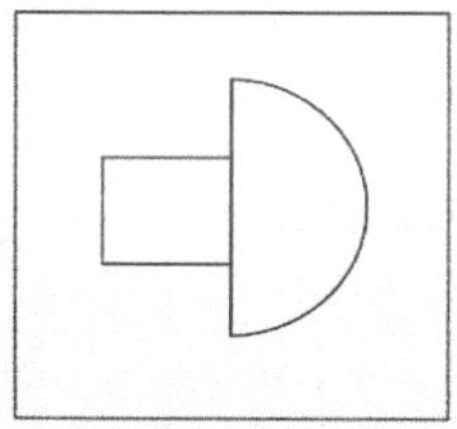
TV AERIAL POINT	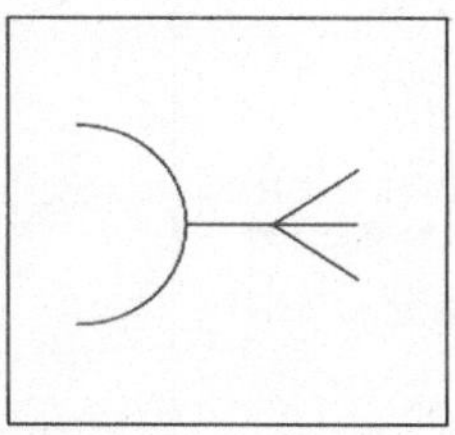	THERMOSTAT	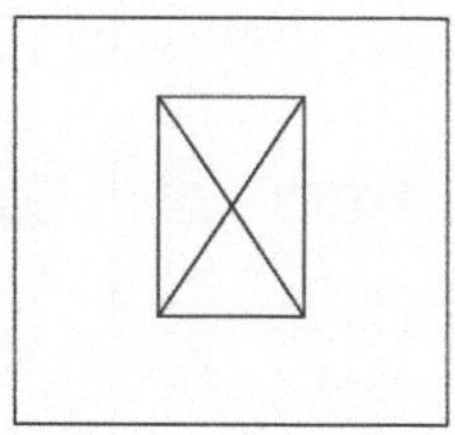
IMMERSION HEATER AND THERMOSTAT	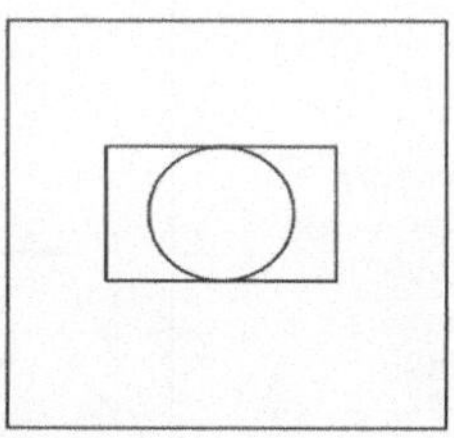	SECURITY ALARM	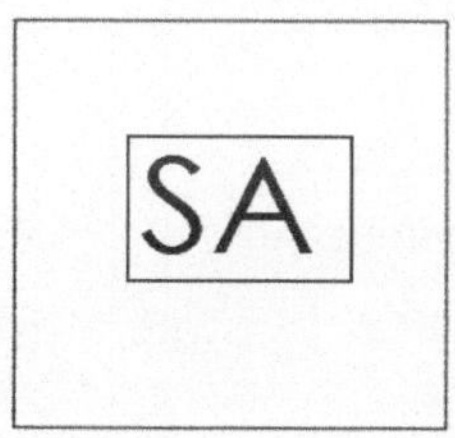

ELEMENT	SYMBOL	ELEMENT	SYMBOL
SWITCHED OUTLET	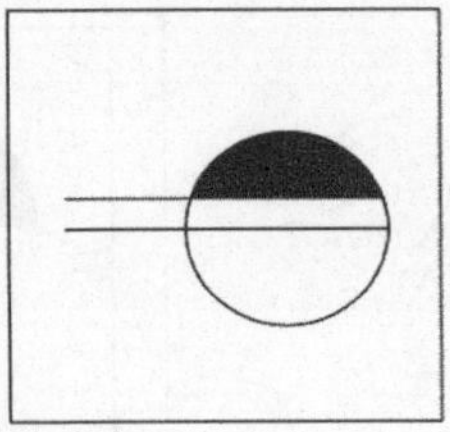	DOOR BELL	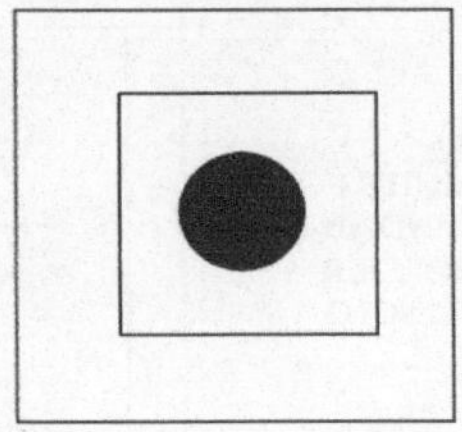
TV OUTLET	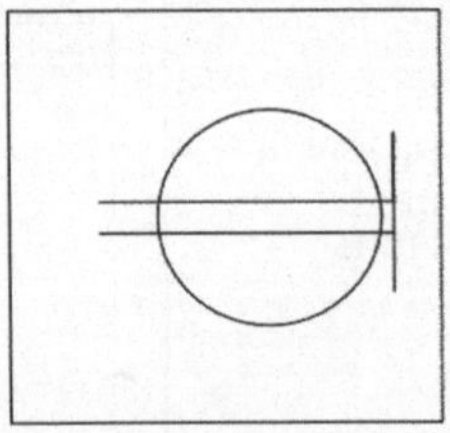	THERMOSTAT	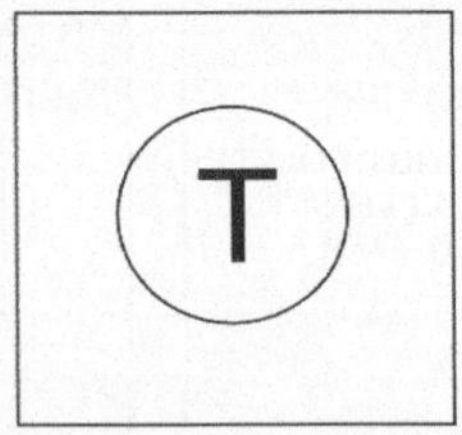
220v OUTLET	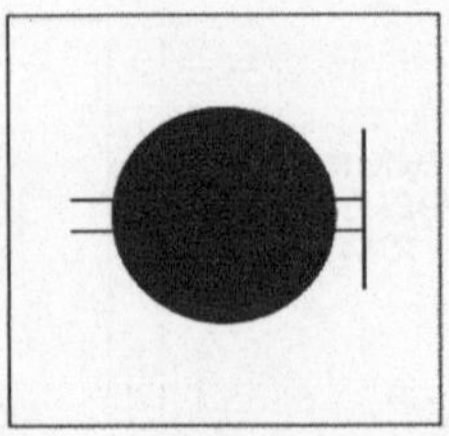	BRASS FLOOR RECEPTACLE	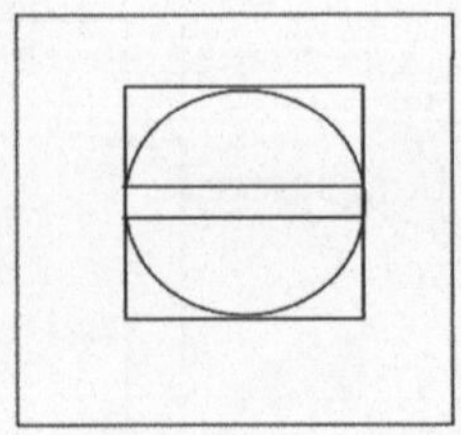
SWITCH (SINGLE)	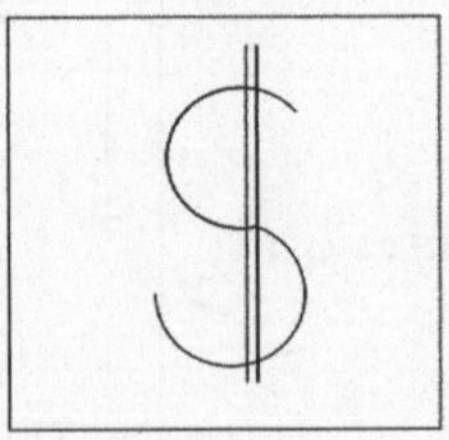	WALL LIGHT/ SCONCE	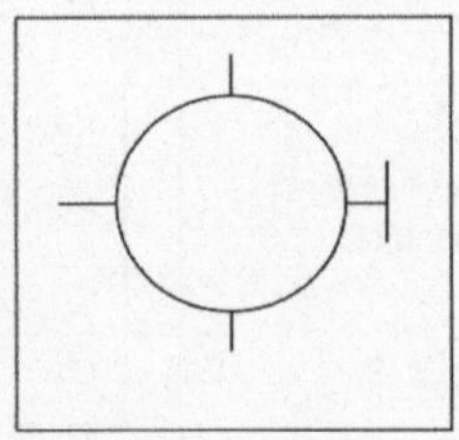
PHONE JACK	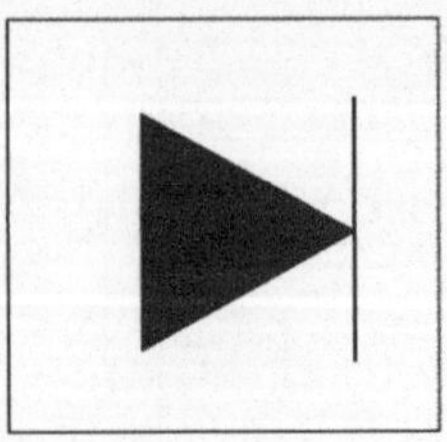	2 CORE SPEAKER CABLE	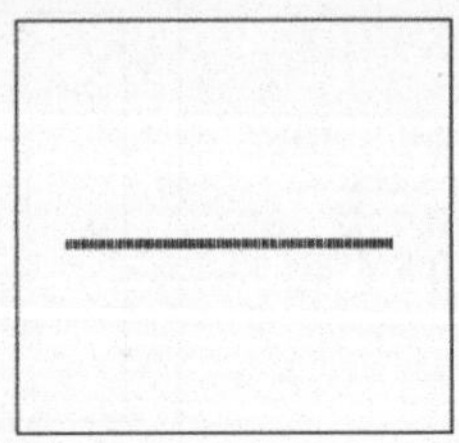

ELEMENT	SYMBOL	ELEMENT	SYMBOL
CONSUMER UNIT		COMBINED HEAT DETECTOR AND FIRE ALARM SOUNDER	
TELEPHONE POINT	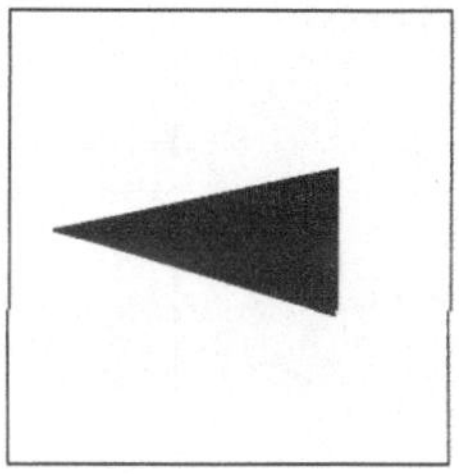	COMBINED SMOKE/HEAT DETECTOR AND FIRE *ALARM* SOUNDER	
CAT 5 or 6 DATA SOCKET		AUDIO SPEAKER	
COOKER SWITCH	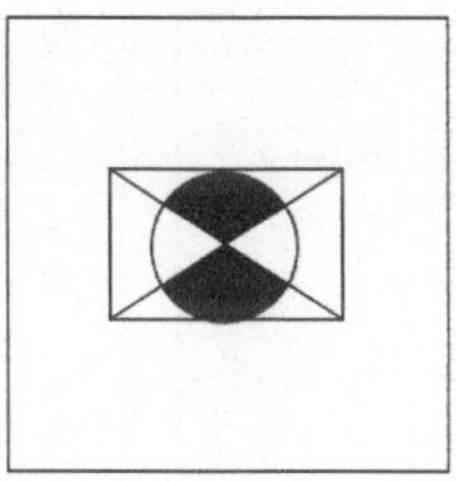	RADIATOR	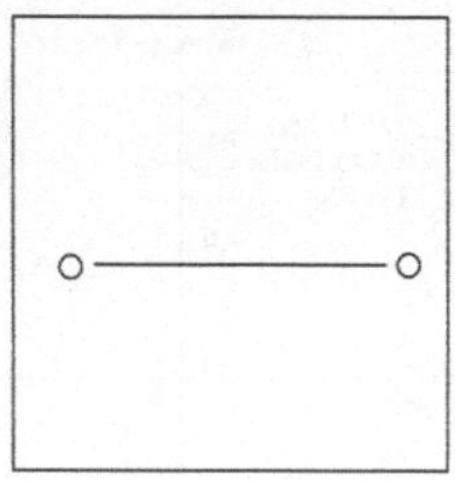

2.5 PLUMBING SYMBOLS

Plumbing fixtures are shown with shapes with an abreivation inside the symbols indicating plumbing symbols.

Table 2.7: Plumbing symbols

ELEMENT	SYMBOL	ELEMENT	SYMBOL
INDIAN TYPE WATER CLOSET	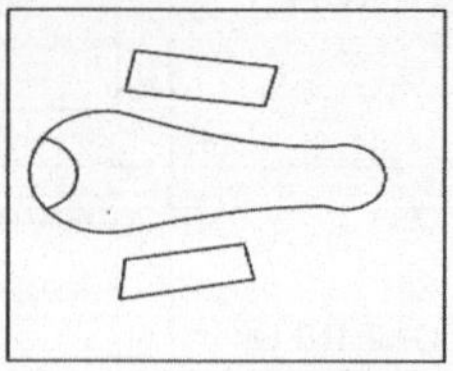	WATER CLOSET WITH FLUSHING SYSTEM	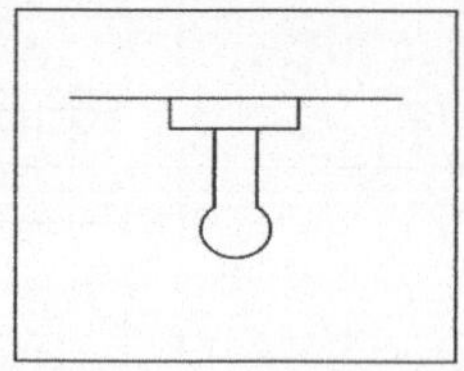
EUROPEAN WATER CLOSET	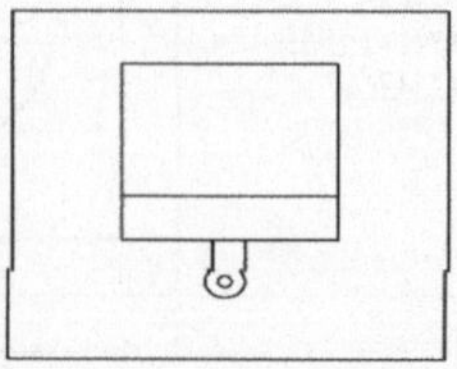	WATER CLOSET WITHOUT FLUSHING SYSTEM	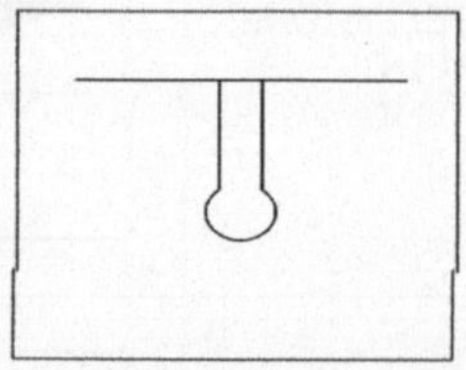
URINAL	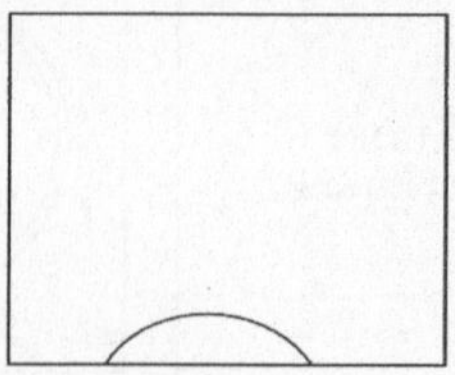	BATH TUB	
URINAL CORNER HUNG	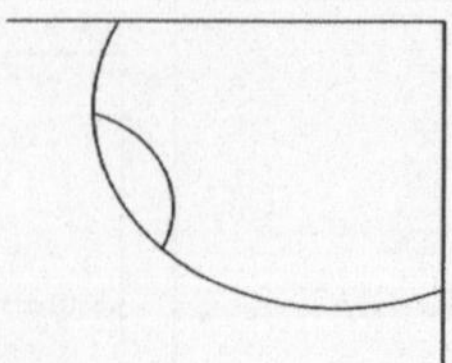	SHOWER	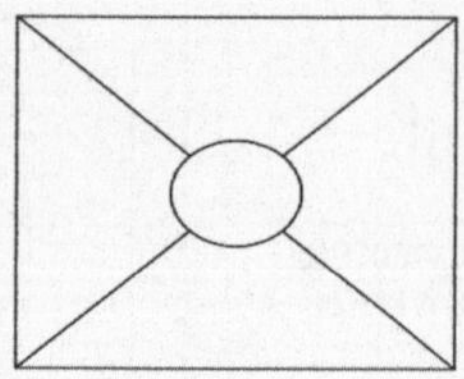
URINAL STALL	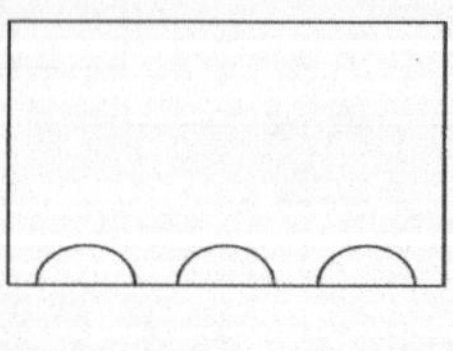	SHOWER HEAD	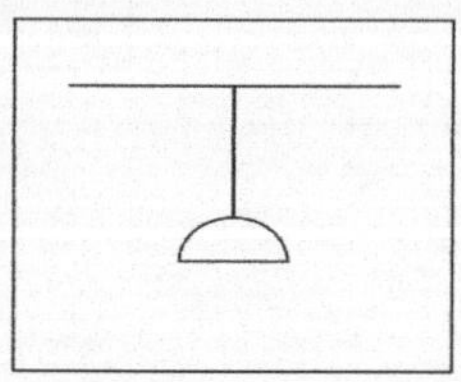

2.6 REBAR DRAWINGS

2.6.1 Footing

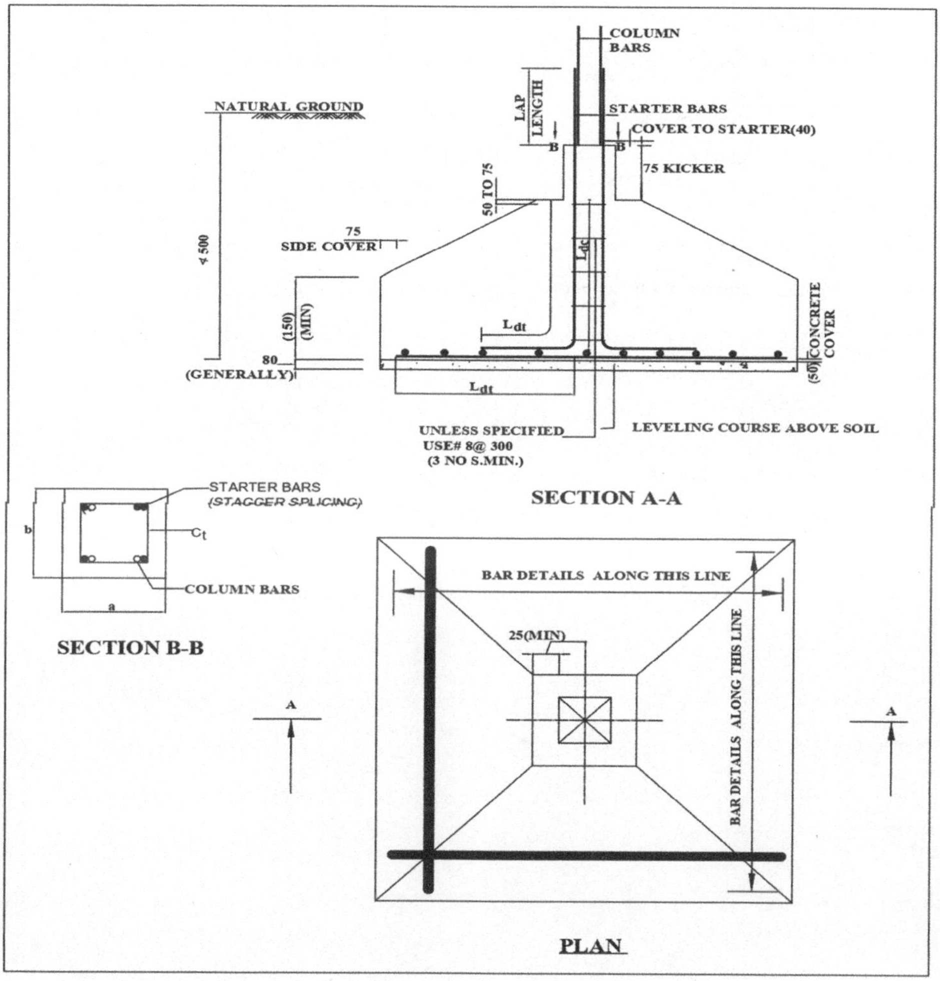

Fig. 2.8: Reinforcement detailing of isolated footing

2.6.2 Column

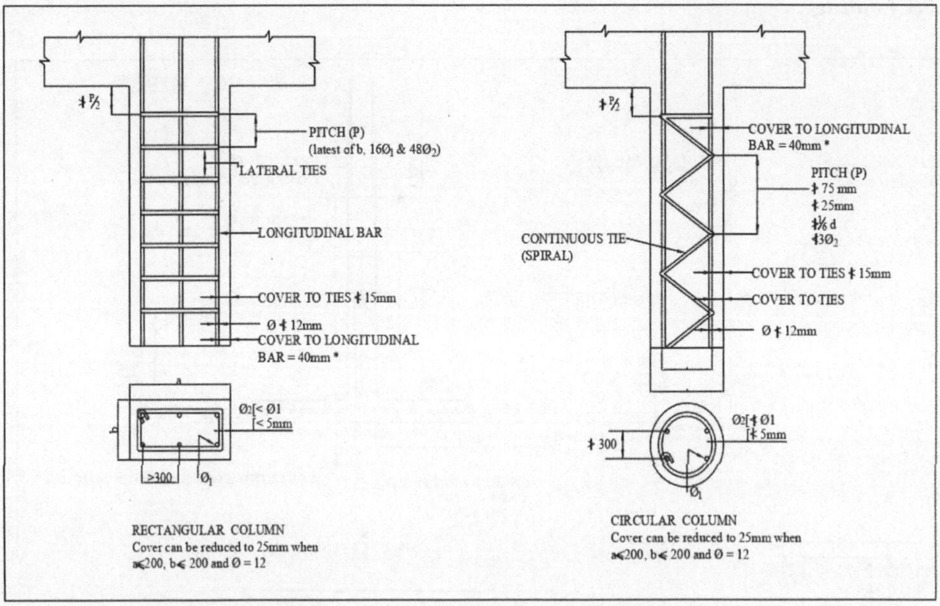

Fig.2.9: Reinforcement detailing of a column

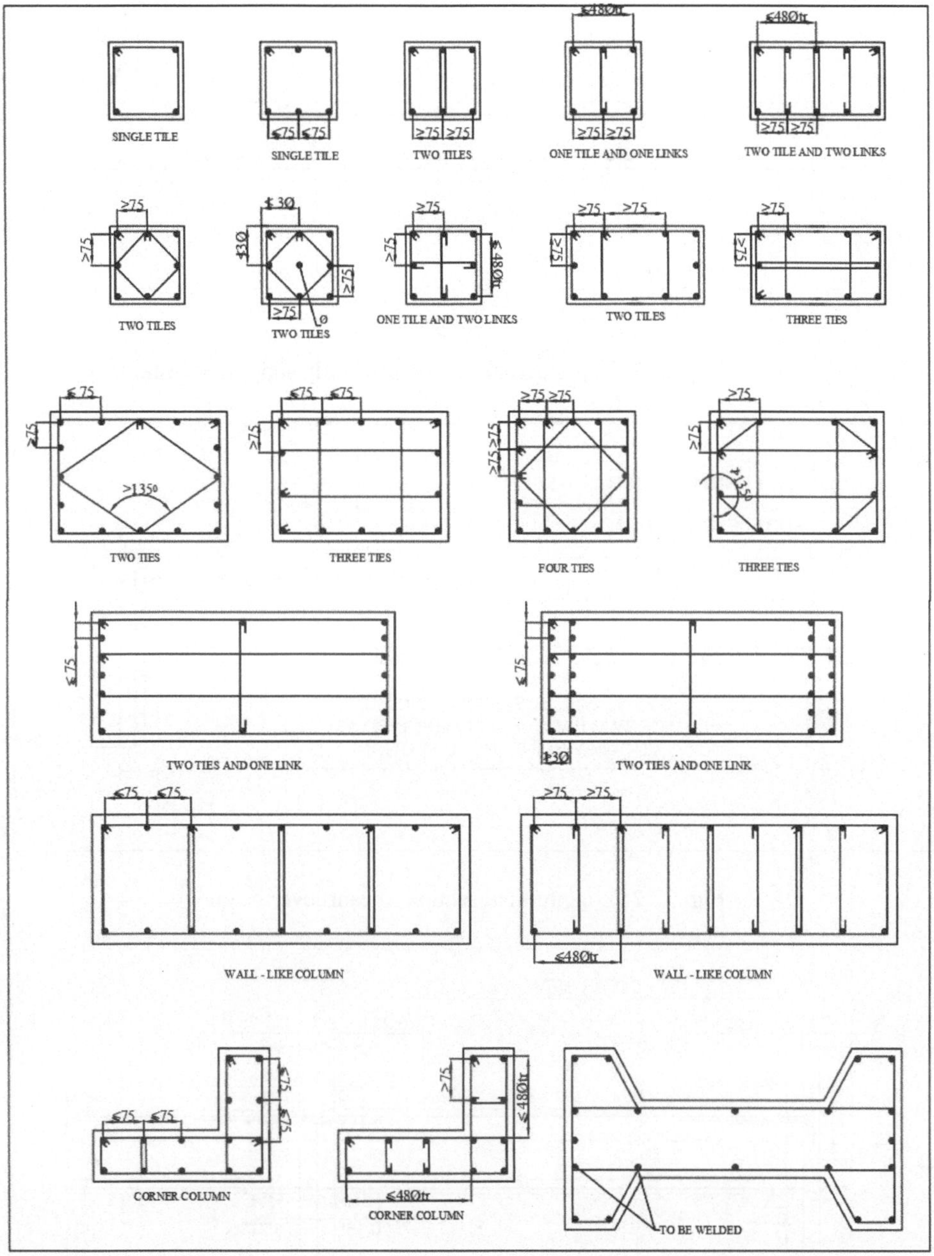

Fig. 2.10: Cross sections of columns and reinforcement detailing

2.6.3 Beam

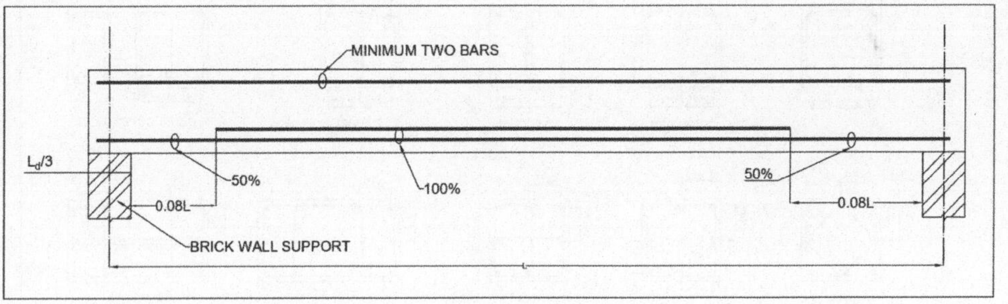

Fig. 2.11: Longitudinal section of simply supported beam

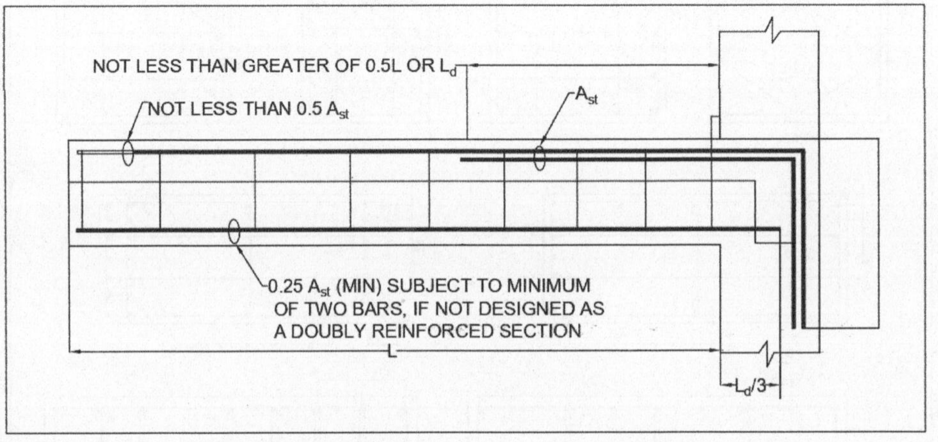

Fig. 2.12: Longitudinal section of cantilever beam

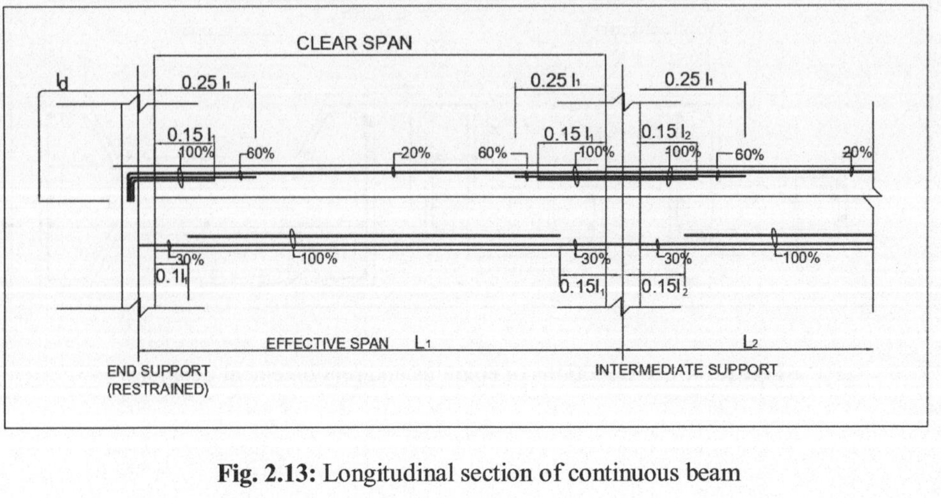

Fig. 2.13: Longitudinal section of continuous beam

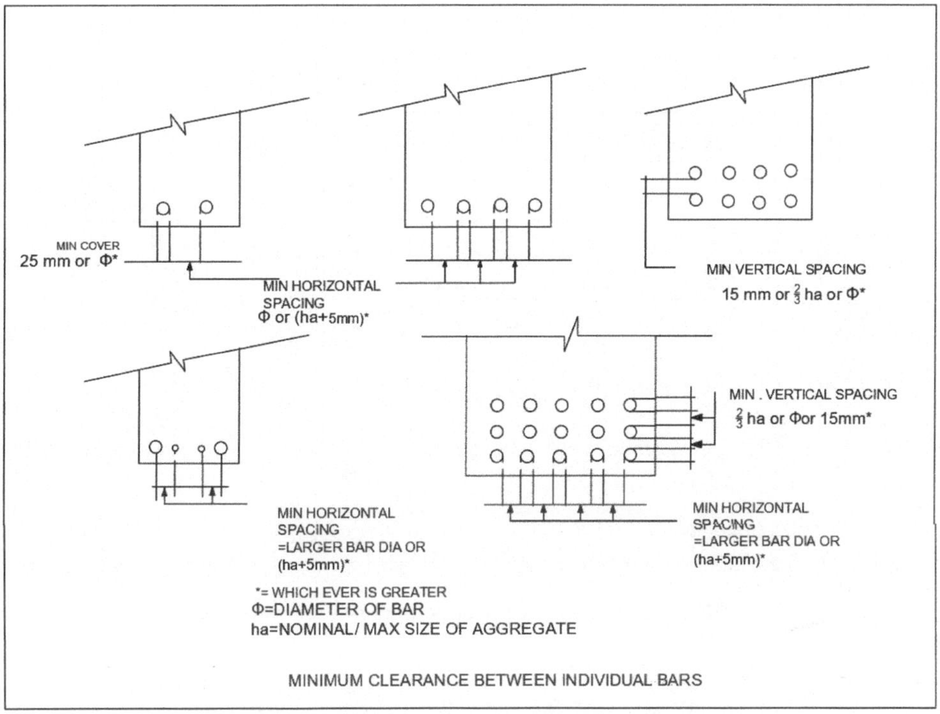

Fig. 2.14: Horizontal and vertical spacing of reinforcement bars

2.6.4 Slab

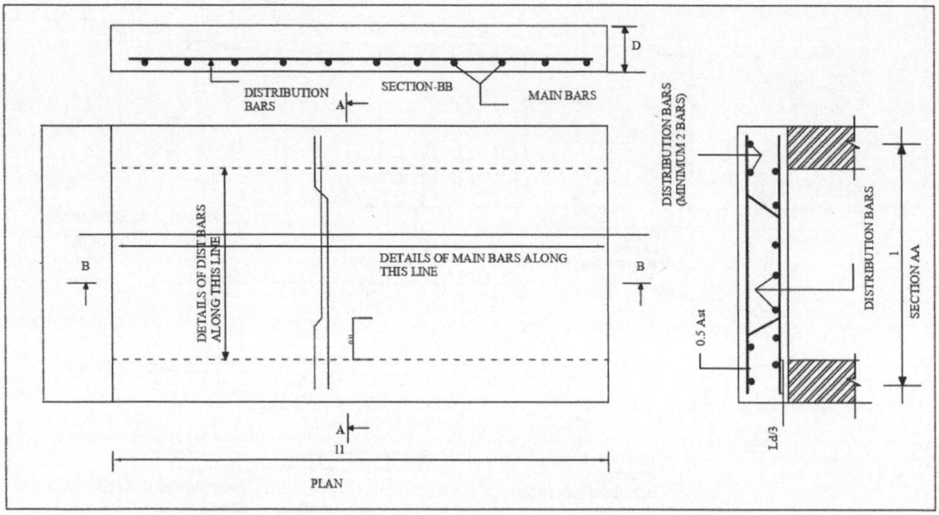

Fig. 2.15: Reinforcement detailing of one-way slab

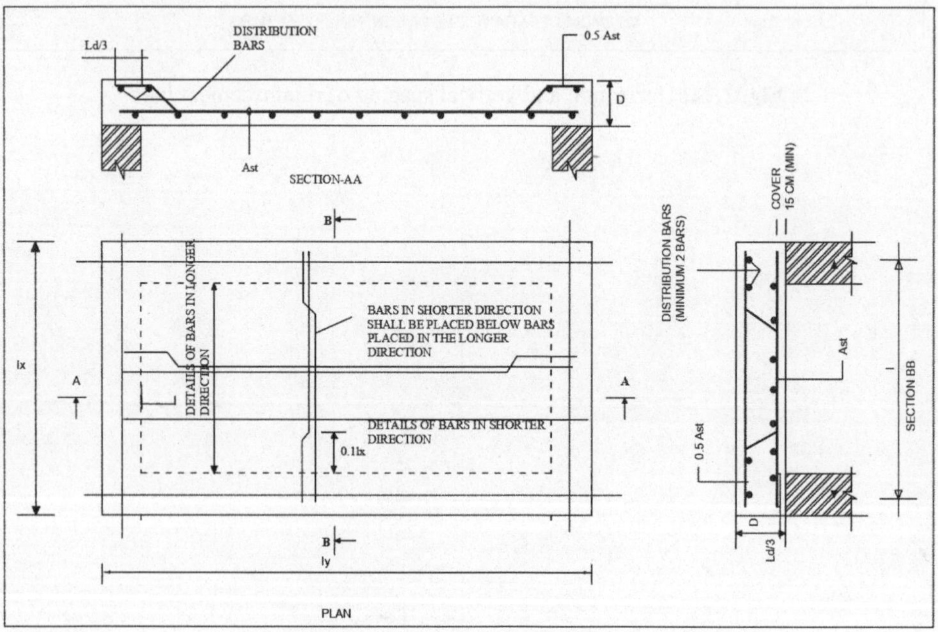

Fig. 2.16: Reinforcement detailing of two-way slab

2.7 STRUCTURAL STEEL FABRICATION AND CONNECTIONS DRAWING SYMBOLS

DESCRIPTION	METHOD OF DRAWING	
	View	Section
Rivet general	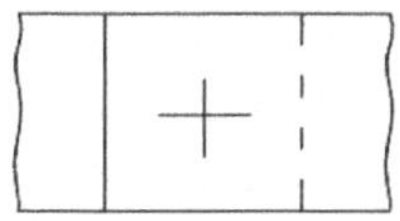	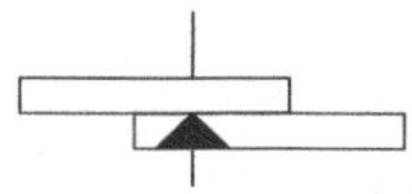
Rivet countersunk on back side	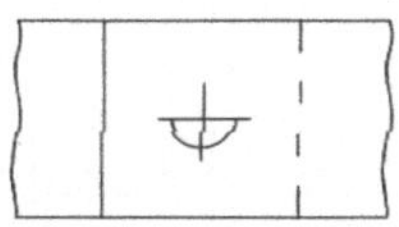	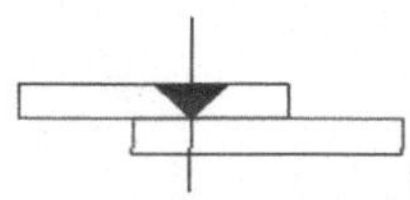
Rivet countersunk on front side	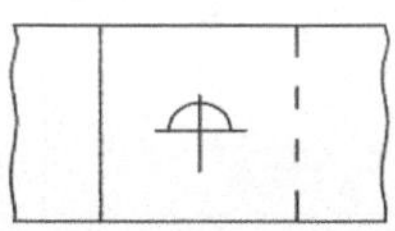	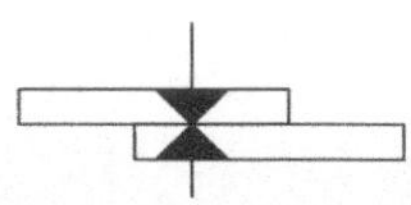
Rivet countersunk on front and back side	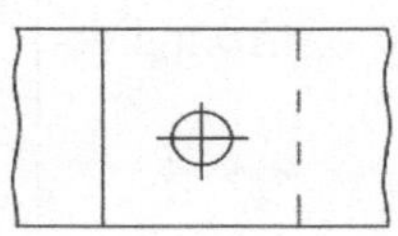	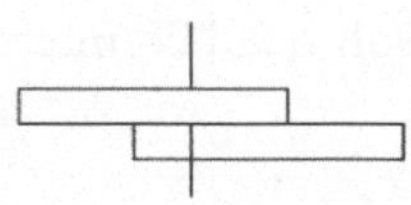
Bolt general	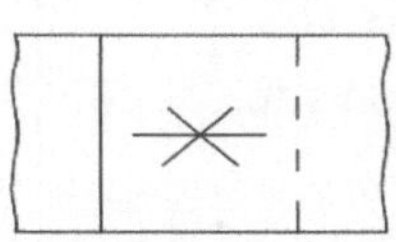	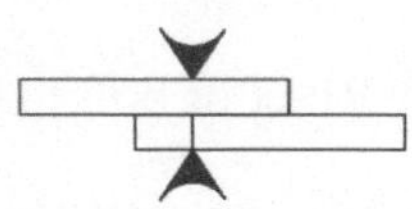

DESCRIPTION	METHOD OF DRAWING	
	View	Section
Bolt to distinguish from the revit	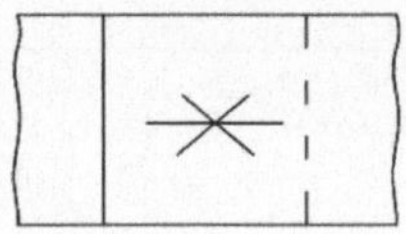	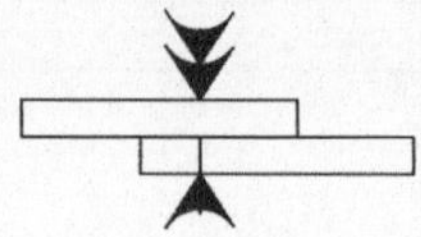
Bolt,place of nut indicated  NUT	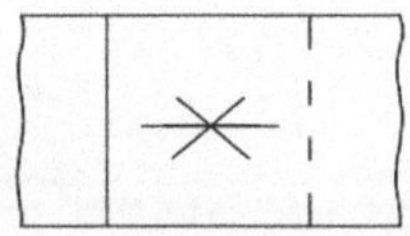	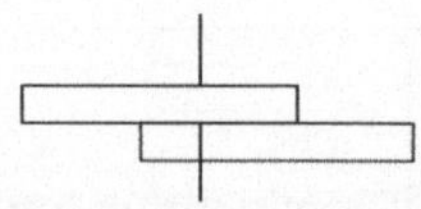
Bolt head countersunk on back side	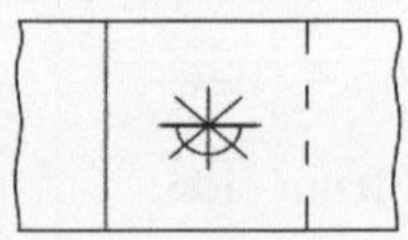	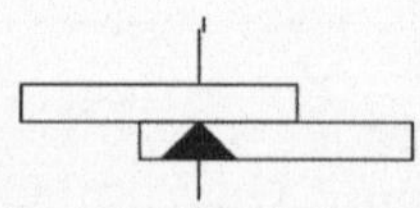
Bolt head countersunk on front side	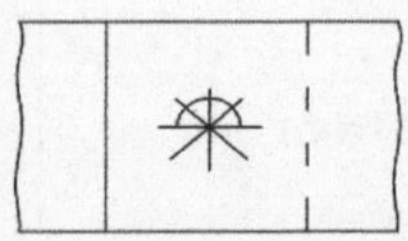	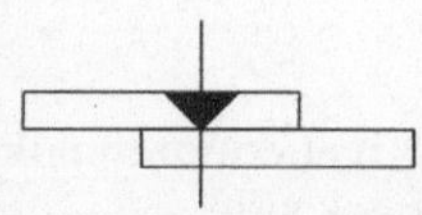
Rivet or bolt to fit on site		

Fig. 2.17: Rivet/Bolted Symbols

FORM OF WELD	SECTION	SYMBOL
FILLET	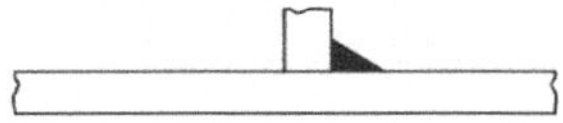	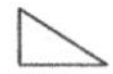
SQUARE BUTT	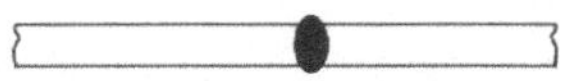	
SINGLE-V-BUTT	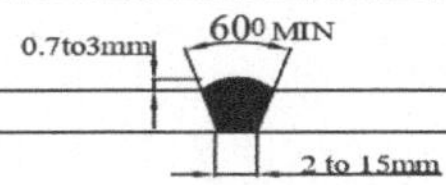	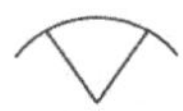
DOUBLE-V-BUTT	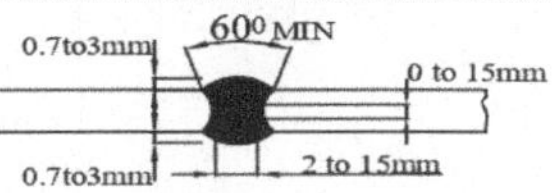	
SINGLE-U-BUTT		
DOUBLE-U-BUTT		
SINGLE-BEVEL-BUTT	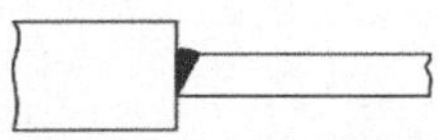	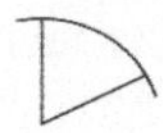
DOUBLE-BEVEL-BUTT	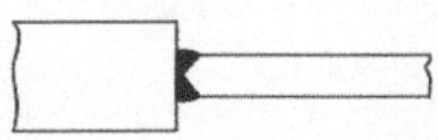	
SINGLE-J-JOINT	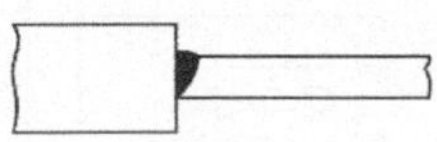	
DOUBLE-J-JOINT	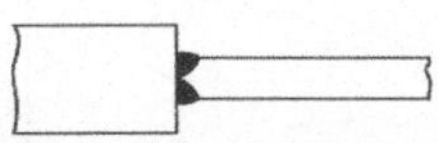	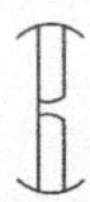

FORM OF WELD	SECTION	DRAWINGS REPRESENTATION
FLUSH BUTT		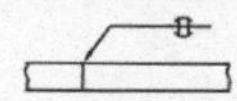 Flush square butt
MACHINE BUTT		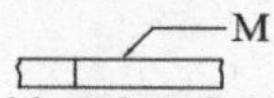 Double V-butt finish flush on another side
GRINDING FINISH	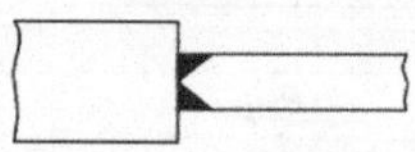	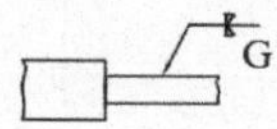 Double bevel butt finished flush on Arrow mark on one section only indicates the section prepared for joint
CONVEX FILLET	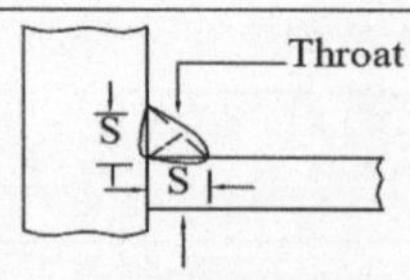	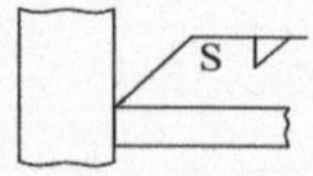
CONCAVE FILLET USING FIELD WELD	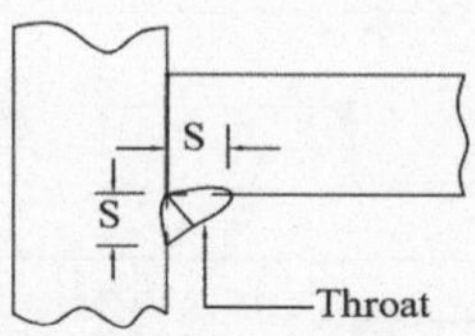	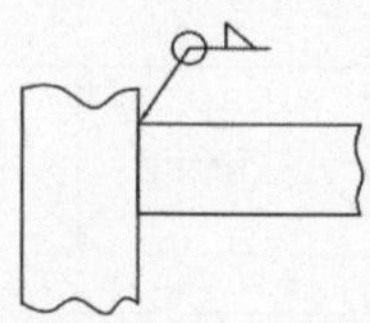 Black circle indicates field weld
WELD ALL ROUND	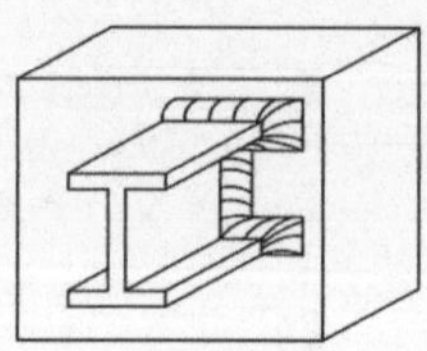	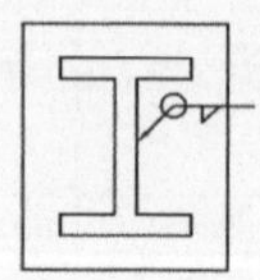 Circle indicates fillet weld all round the section

Fig. 2.18: Welding Symbols

2.7.1 Fabrication Drawings for Rivets/Bolts

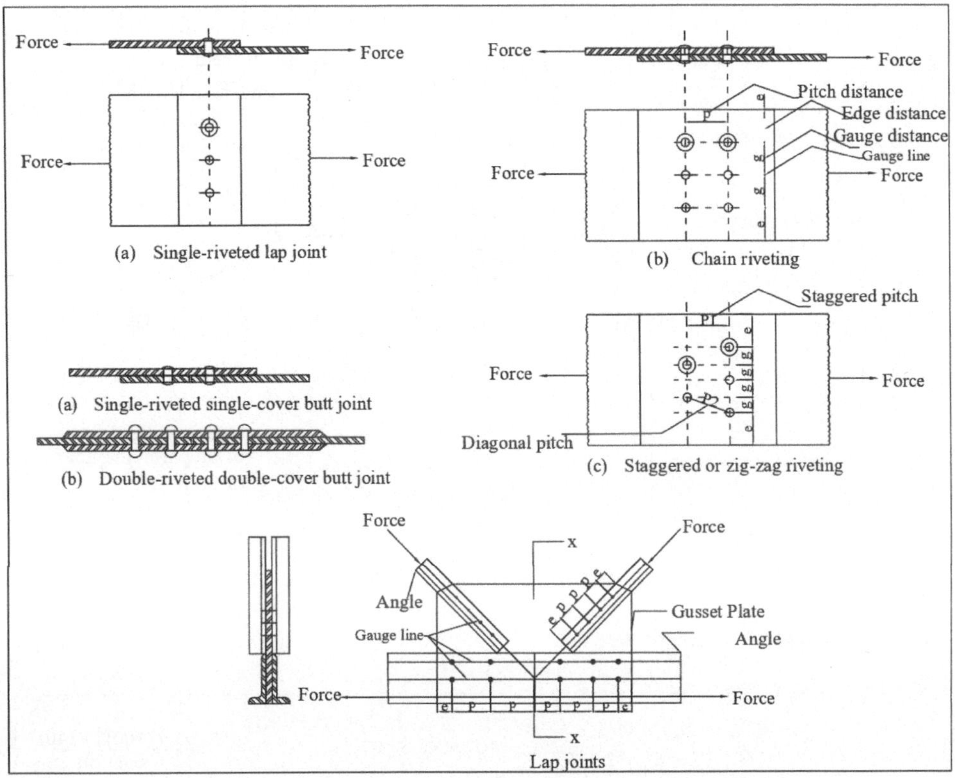

Fig.2.19: Fabrication Drawing for Rivet/Bolt

2.7.2 Fabrication Drawings for Welding

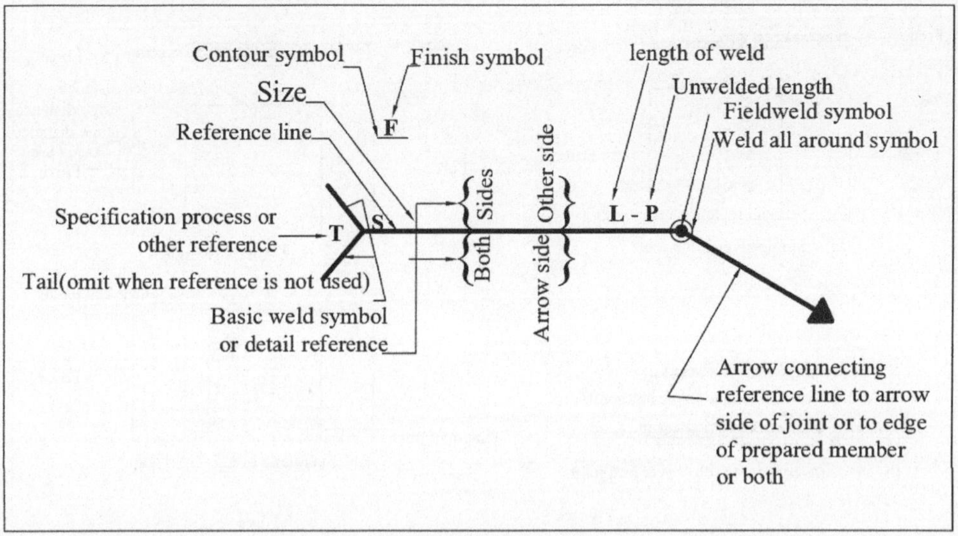

Fig. 2.20: Fabrication drawing for welded connections

Example for welded connections

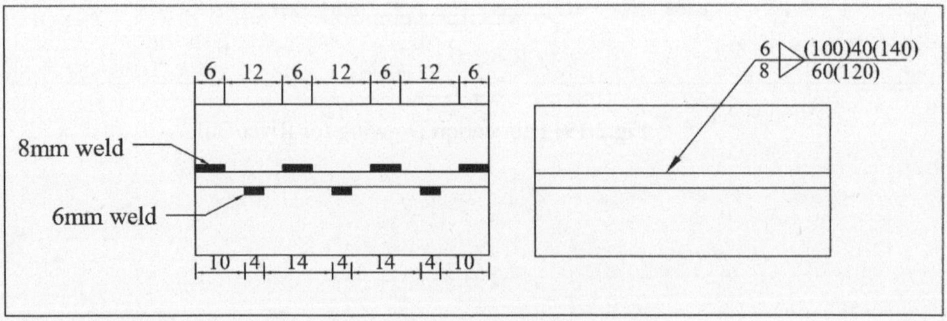

2.8 DIMENSIONING STANDARDS

When the object or an element is drawn, the sizes of all the parts are to be shown using dimensions. The various definitions to be remembered for dimensioning as object are

1. **Dimension:** The numerical value of the individual element in an object.
2. **Basic Dimension:** It refers to the theoretically precise values that specify a portion or features form, size, orientation or position.
3. **Center Mark:** It represents the center of a circle, semicircle, or hole, represented with '+' symbol.
4. **Leader Line:** A leader line on a drawing is the line that joins a dimension, note, or a label to a feature.
5. **Extension Line:** A line that goes from a point on a part or feature to the beginning of a dimension line, arrow or leader line.
6. **Arrowhead:** A dimension line or a leader line is indicated by an arrowhead at the start and end point in technical drawings.
7. **Diameter Symbol:** To indicate that a measurement is for the diameter of a circle or cylindrical item, represented with a symbol 'Φ'.
8. **Radius Symbol:** To indicate that a measurement is for the radius of a circle, represented with the letter 'R'.

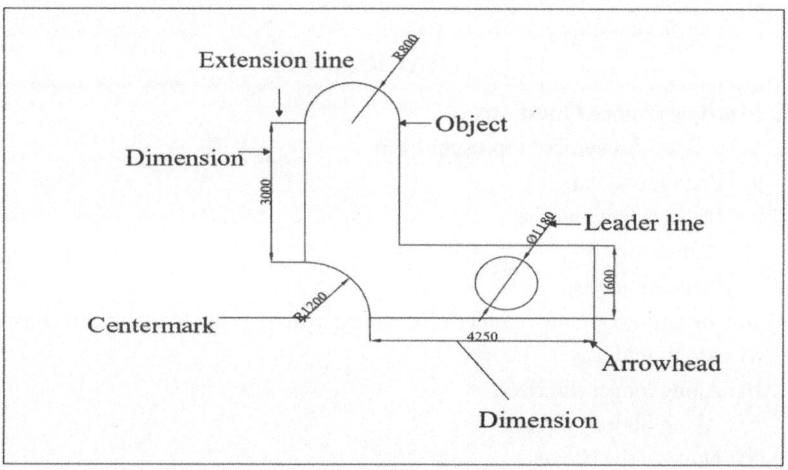

Fig. 2.21: Dimensioning standards

UNIT SUMMARY

This unit makes you learn more about standardizing drawings and the points to remember are

- The sign conventions should be remembered and shown exactly in the drawing where the specific material is used.
- A slight variation in depicting a sign can lead to a change in the behaviour of the structure after construction.
- The significance of the symbol is exactly required for rebar scheduling
- The connections length, cutting length, offsets, lapping of bars, covers, and general notes play an important role in improving the efficiency during the construction phase
- All the longitudinal sections, cross sections, and plans support length, and spacing between rebar should be followed as per specific codes. Violating codal provisions leads to distress in the elements.
- Technical and execution compliances should be clearly studied before preparing rebar drawings.

EXERCISES

A. Multiple Choice Questions

1. What does this symbol represent 'Φ'?
   a) Plain round bar
   b) Plain rectangular bar
   c) Round bar
   d) Plain Square bar
2. In which direction the main reinforcement will be provided for a one-way slab?
   a) Along mid way
   b) Along longer direction
   c) Along Shorter direction
   d) None of the above
3. 'St' symbol represents
   a) Straight bend
   b) Top bar
   c) Stretched bar
   d) Straight bar

4. Minimum clear cover for footing
   a) 30mm
   b) 25mm
   c) 50mm
   d) 65mm
5. Which one of these are not topography symbols
   a) Sanitary sewers.
   b) Pump
   c) Grass
   d) Gas line
6. The minimum diameter of longitudinal bar in column is
   a) 10mm
   b) 14mm
   c) 12mm
   d) 20mm
7. Which of these are not forms of rivet/bolted?
   a) Square butt
   b) Grinding Finish
   c) Single-J-Joint
   d) All the above

Answers

1. a 2. c 3. d 4. c 5. b 6. c 7. d

B. Short and Long Answer Type Question;

2.1 List out and draw sign conventions for some of the materials.

2.2 What are the different type of reinforcements used in construction?

2.3 Describe about dimension standards.

2.4 Draw an isolated footing of depth 600mm and dimensions of 1200 X 1200mm, also show the reinforcement detailing .Assume the necessary data.

2.5 What are Plumbing symbols and list any five symbols.

2.6 List out different types of doors and windows used in residential buildings.

KNOW MORE

- *Connections:* Developing connections of various structural elements needs sound knowledge and is one of the toughest jobs for an engineer learning in-depth tools and can make it effective.
- *Developing Sections:* The student needs to understand the importance of each unit of a drawing, visualize the exact need of the various sections, and develop the sections required.
- *Plan to Execute:* Learning individual sections of all the elements, material specifications, signs, and symbols the student needs to develop all the essential drawing sections with the help of drawing tools.

REFERENCES AND SUGGESTED READINGS

1. S.C. Rangwala. Engineering Materials, 44th ed.; Charotar Publications, 2024.
2. N. Subramanian. Building Materials, Oxford Publications, 2019.
3. IS SP 34 (1987): Handbook on Concrete Reinforcement and Detailing.
4. L.S. Negi. Design of Steel Structures, 2nd ed.; 2018.

Dynamic QR Code for Further Reading

3 MASONRY BONDS

UNIT SPECIFICS

The following aspects are discussed in this module.

- Masonry bond
- Sizes of brick
- Basic Definitions in brick masonry
- Bonds in brick masonry
- Corner walls and cross walls

RATIONALE

This section explains the various types of masonry bonds that are commonly used in the construction of a strong and aesthetically pleasing wall. The method for laying the bricks in each course to make the wall stable under various loading conditions is also explained. Additionally, the drawings of wall sections at various junctions and with varying thicknesses are described with neat sketches.

PRE-REQUISITES

- Types and sizes of bricks
- Terms in brickwork
- Types of walls

UNIT OUTCOMES

The following are the outcomes of this unit:

U3-O1: Brief understanding of masonry and bond

U3-O2: Introductory specification of brick masonry

U3-O3: Understanding and drawing of English bond and Flemish bond

U3-O4: Understanding and drawing of corner wall and cross wall

Unit- 3 Outcomes	EXPECTED MAPPING WITH COURSE OUTCOMES (1-Weak correlation; 2-Medium correlation; 3-Strong correlation)				
	CO-1	**CO-2**	**CO-3**	**CO-4**	**CO-5**
U3-O1	3	3	3	1	2
U3-O2	2	1	3	1	1
U3-O3	3	3	3	2	2
U3-O4	3	3	3	2	2

3.1 Masonry Bond

The construction of a monolithic unit by bonding several individual units like bricks and stones is called as masonry. Bond is the systematically arrangement of stones and bricks to ensure strength and stability to the wall.

The masonry bond is to ascertain firm stability and interlocking between all the building units such that the wall can carry the maximum possible load without failure.

Masonry can be classified based on materials used as brick masonry, stone masonry, concrete masonry, composite masonry, and reinforced brick masonry. The most commonly used masonry is brick masonry, which is economical and widely accepted due to the availability of raw material (clay) in abundance.

Bond types in brick masonry wall construction are classified according to the laying and bonding styles of bricks in walls. Brick masonry bonds are formed by mortar filling between layers of bricks, grooves formed when bricks are set near each other, and layers in walls.

3.2 Size of Brick

After careful consideration, the Bureau of Indian Standards standardized the standard size of a modular brick as 200 x 100 x 100mm. The actual dimensions of a standard brick are 190 x 90 x 90mm. The mortar joint is 10mm thick. Therefore, 200 x 100 x 100mm is considered to be the standard brick size.

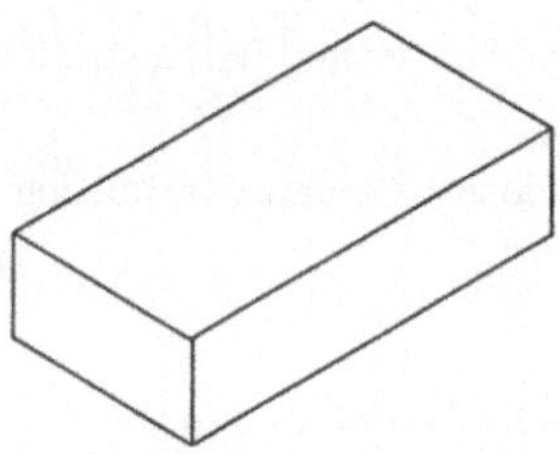

Fig. 3. 1: Full Brick

3.3 Basic Definitions in Masonry Bond

a) Stretcher: Bricks that are placed with their length parallel to the front face, or direction of a wall are called stretchers. Stretcher courses refer to courses that contain stretchers.

b) Header: Bricks that are placed with their width or breadth parallel to the front, face, or direction of a wall are called headers. Header courses refer to courses that contain headers.

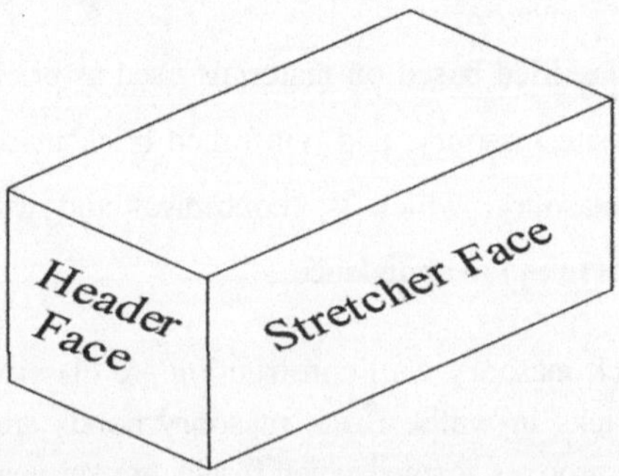

Fig. 3. 2: Stretcher and Header Face

c) Closer: A closer is a piece of brick that is used to seal the joint at the end of brick courses. Closers are classified into four types.

(i) Queen Closer: The bricks are divided into two equal pieces by cutting them longitudinally. It can also be constructed using two quarter bricks or a quarter closer.

(ii) King Closer: King closer is made by cutting a triangle-shaped section of the brick so that the adjacent cut faces create half a header and half a stretcher.

(iii) Beveled Closer: A beveled closer is created by cutting a triangular part half the width but its entire length. It appears as a closer on one face and as a header on the other.

(iv)Mitred Closer: To create a mitred closer, cut a triangle through the width of the brick and make an angle of 45^0 to 60^0 degrees with the length of the brick. Following are the different closers used in brick masonry.

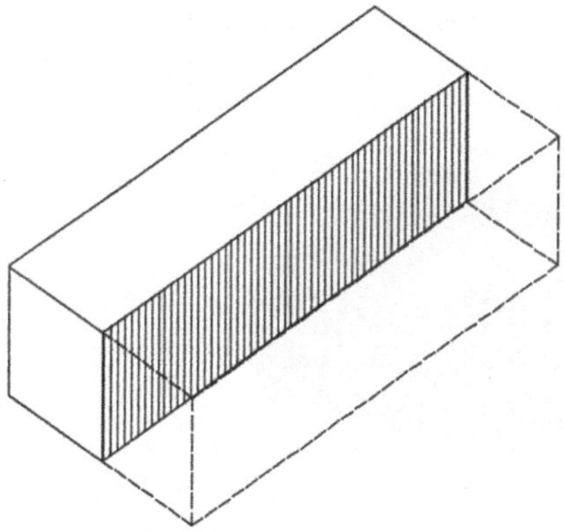

Fig. 3. 3: Queen closer

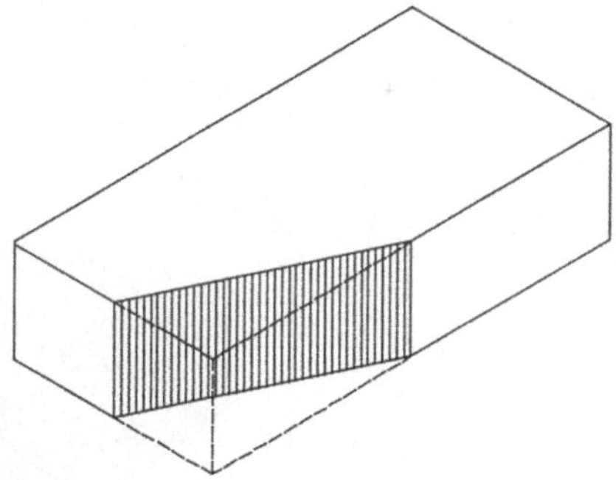

Fig. 3. 4: King closer

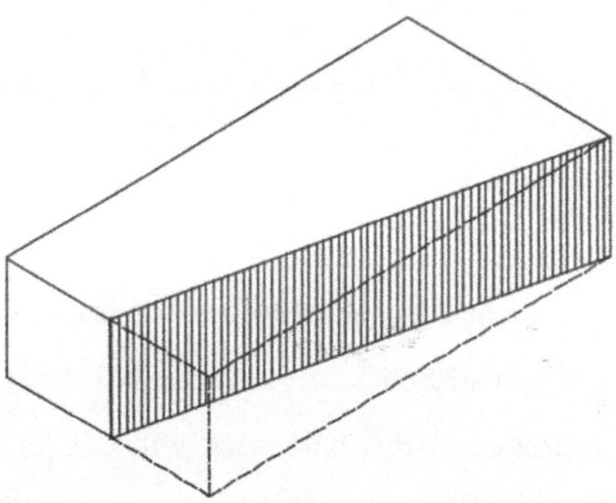

Fig. 3. 5: Beveled closer

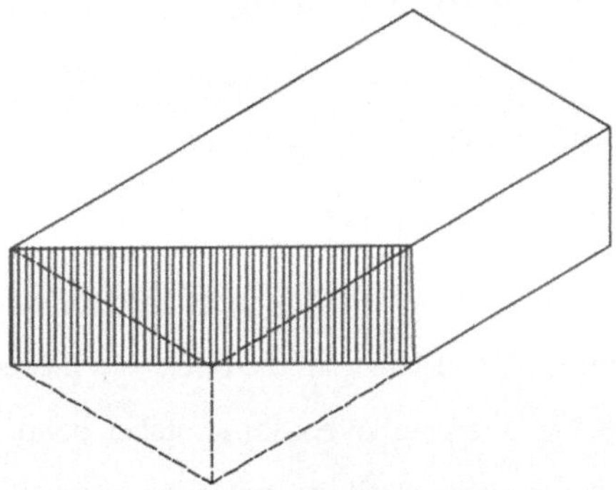

Fig. 3. 6: Mitred closer

d) Bat: It is a part of the brick; it can be half bat, three-quarter bat.

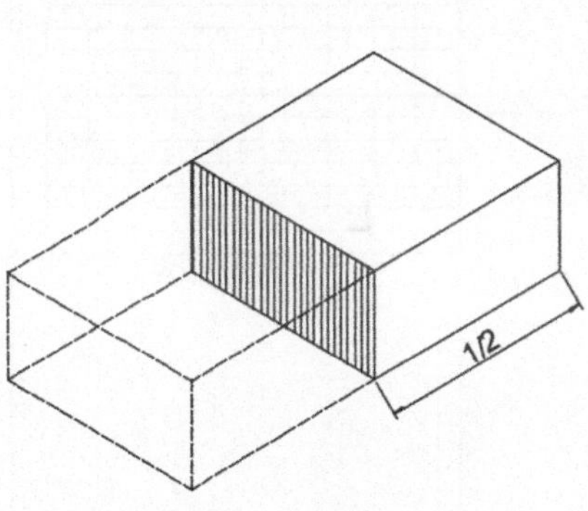

Fig. 3. 7: Half bat

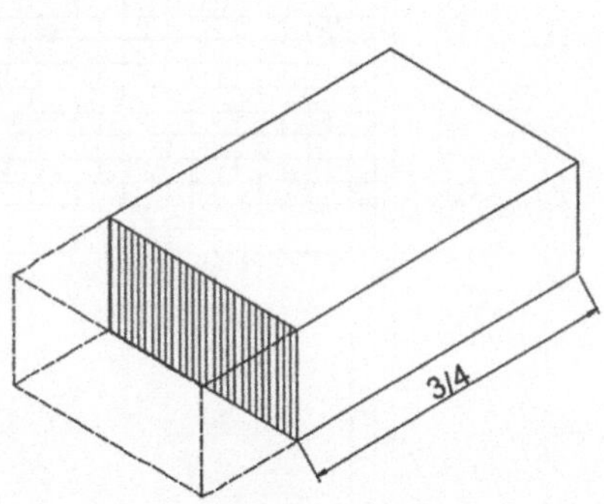

Fig. 3. 8: Three – quarter bat

e) Frog: A frog is a mark that is applied to the face of a brick to create a key for holding the mortar. Its depth ranges from 10 to 20mm.

Fig. 3. 9: Frog

3.4 Bonds in Brick Masonry

The most commonly used bonds in brick masonry are Stretcher bond, Header bond, English bond and Flemish bond

3.4.1 Stretcher Bond

The types of brick walls used for walls with a thickness not greater than one brick wall. In this type of bond, all the bricks are placed uniformly in stretcher courses. One half brick is used in alternate layers for stretcher bond, as the presence of this half brick will not let the vertical plane of all the bricks be in one straight line, which avoids the weakest plane. The elevation and plan of the stretcher bond are shown in the Fig. 3.10.

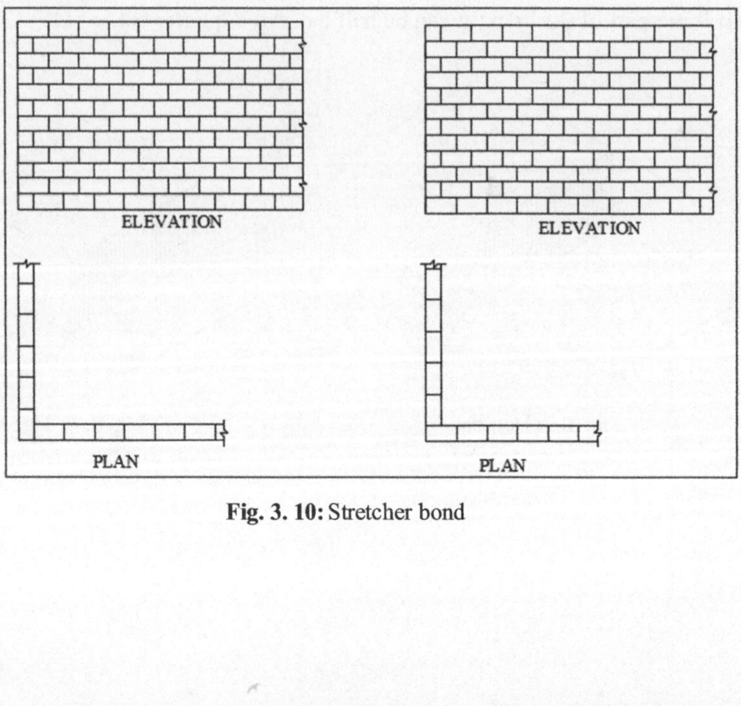

Fig. 3. 10: Stretcher bond

3.4.2 Header Bond

All of the bricks are organized in header courses in header bond. Three-quarter brickbats are used as quoins in each alternate course to keep the overlap at a constant half of the width of the brick. The strength of this bond prevents it from transmitting pressure along the length of the wall. As a result, it is not appropriate for load-bearing walls. The elevation and plan of the header bond are shown in the Fig. 3.11.

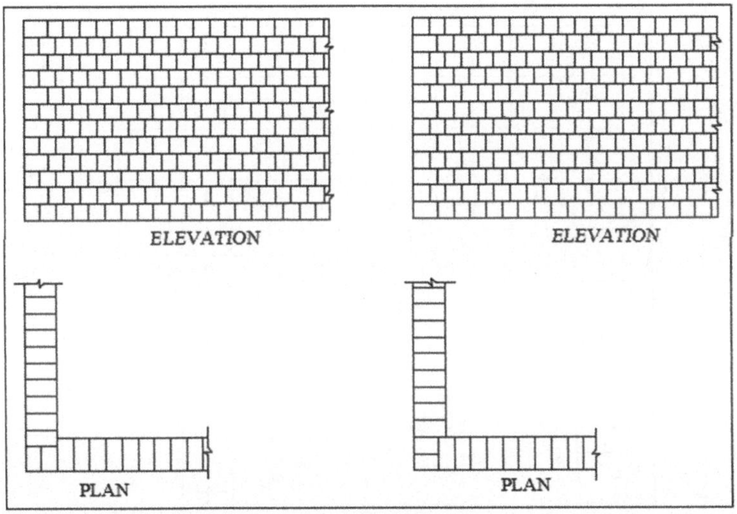

Fig. 3. 11: Header bond

3.4.3 English Bond

The English bond in brick masonry consists of two alternating courses of stretchers and headers, each with one course of stretcher and one course of header above it. Each alternate row is vertically aligned and the headers are placed in the middle of the stretchers in the course below. After the initial header, quoin closer is used at the beginning and end of a wall to break up the continuity of vertical joints. Bricks are divided lengthwise into two pieces and used as quoin closes at corners of brick walls.

Basic Features of the English Bond

1. A queen closer should never begin a header course, as it could be easily displaced. A queen closer is always offered following the quoin header.
2. Each alternate header is centrally placed over a stretcher.
3. Continuous vertical joints are not preferred.

4. The wall has the same appearance on both the front and back faces if the thickness is an even number of half bricks, or one brick or two bricks.
5. The same course will have stretchers on one face and headers on the other face if the wall thickness is an odd number of half bricks (1½ and 2½bricks).
6. Only headers should be placed in the hearting (The core between the facing and backing) of thicker walls.

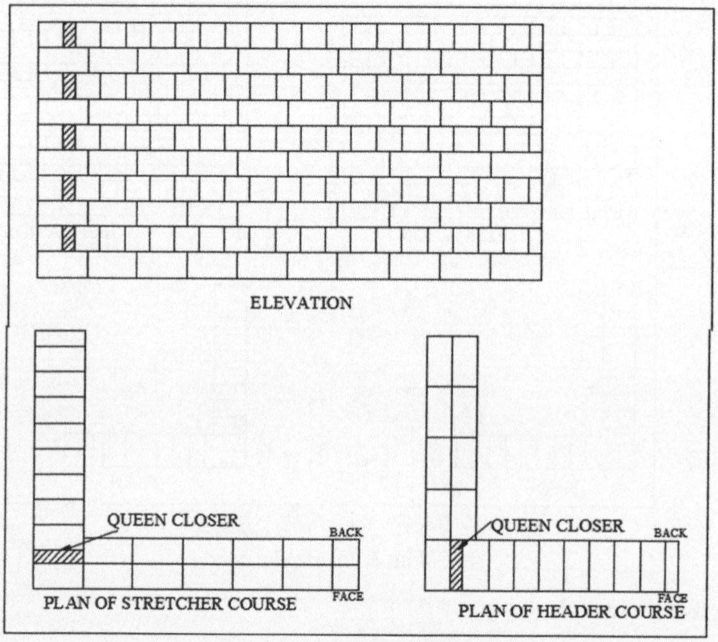

Fig. 3. 12: English bond - 1 Brick wall

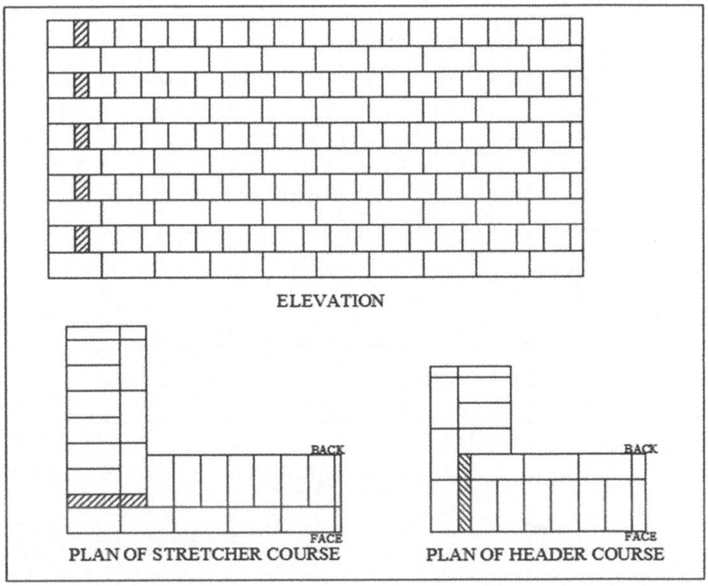

Fig. 3. 13: *English bond - 1 ½ Brick wall*

3.4.4 Flemish Bond

Each course in this bond is made up of alternating headers and stretchers. Every alternate course begins with a header in the corner. To break up the vertical joint's continuity, queen closers are positioned in alternating courses next to the quoin header. Each header is supported centrally by the stretcher beneath it. There are two types of Flemish bond.

(a) Single Flemish bond: have an English bond at its backing but a Flemish bond at its facing. The minimum thickness required for this bond is 1 ½ brick wall.

(b) Double Flemish bond: have a Flemish bond facing and backing.

Basic Features of the Flemish Bond

1. Every course has alternately positioned headers and stretchers
2. In each course, the wall's facing and backing have the same appearance
3. In every alternate course, queen closers are positioned next to the quoin header
4. In walls with thickness equivalent to odd multiples of half bricks, half bats and three-quarter bats are widely utilized.
5. No bats are needed for walls whose thickness is an even multiple of half bricks. A header or stretcher will merge in the same direction on both the front and back faces.

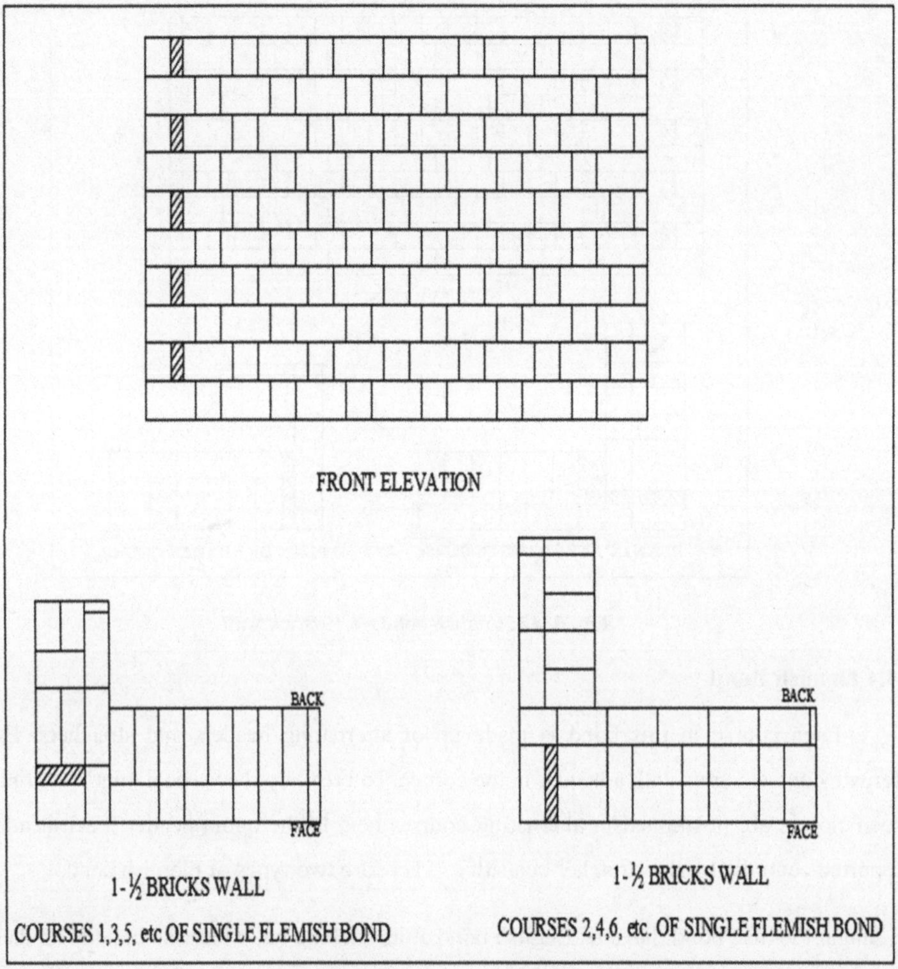

Fig. 3. 14: Single Flemish bond

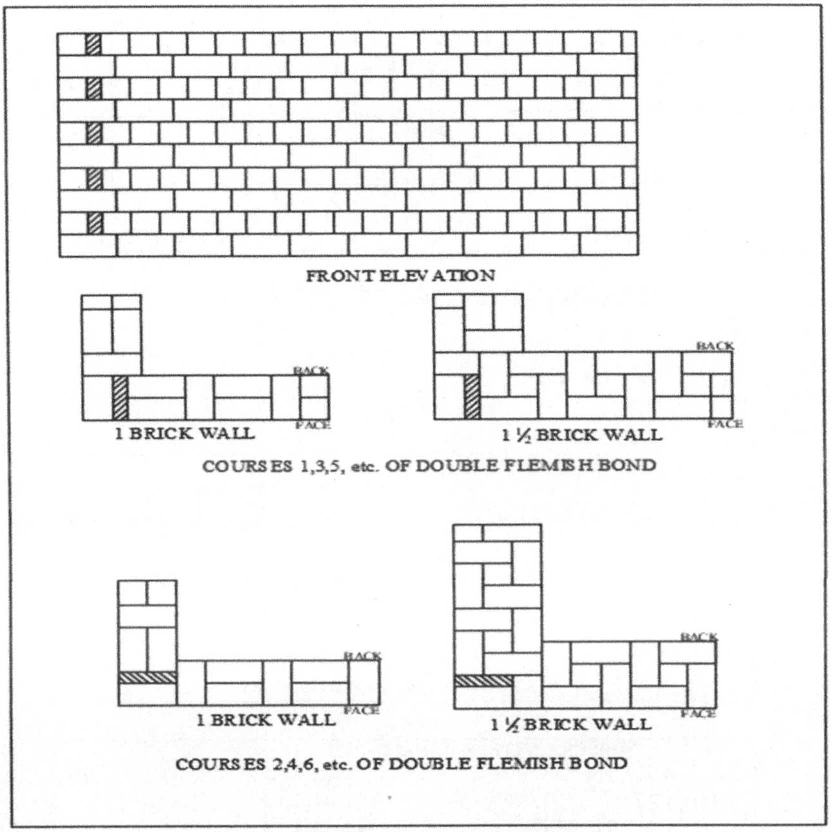

Fig. 3. 15: Double Flemish bond

3.5 CORNER WALL AND CROSS WALL

3.5.1 Corner Wall: Corner walls are observed at the junctions of two different walls or at the corners of a room or building. It is also called as Tee-Junction. This type of junctions is observed in the exterior part of the building.

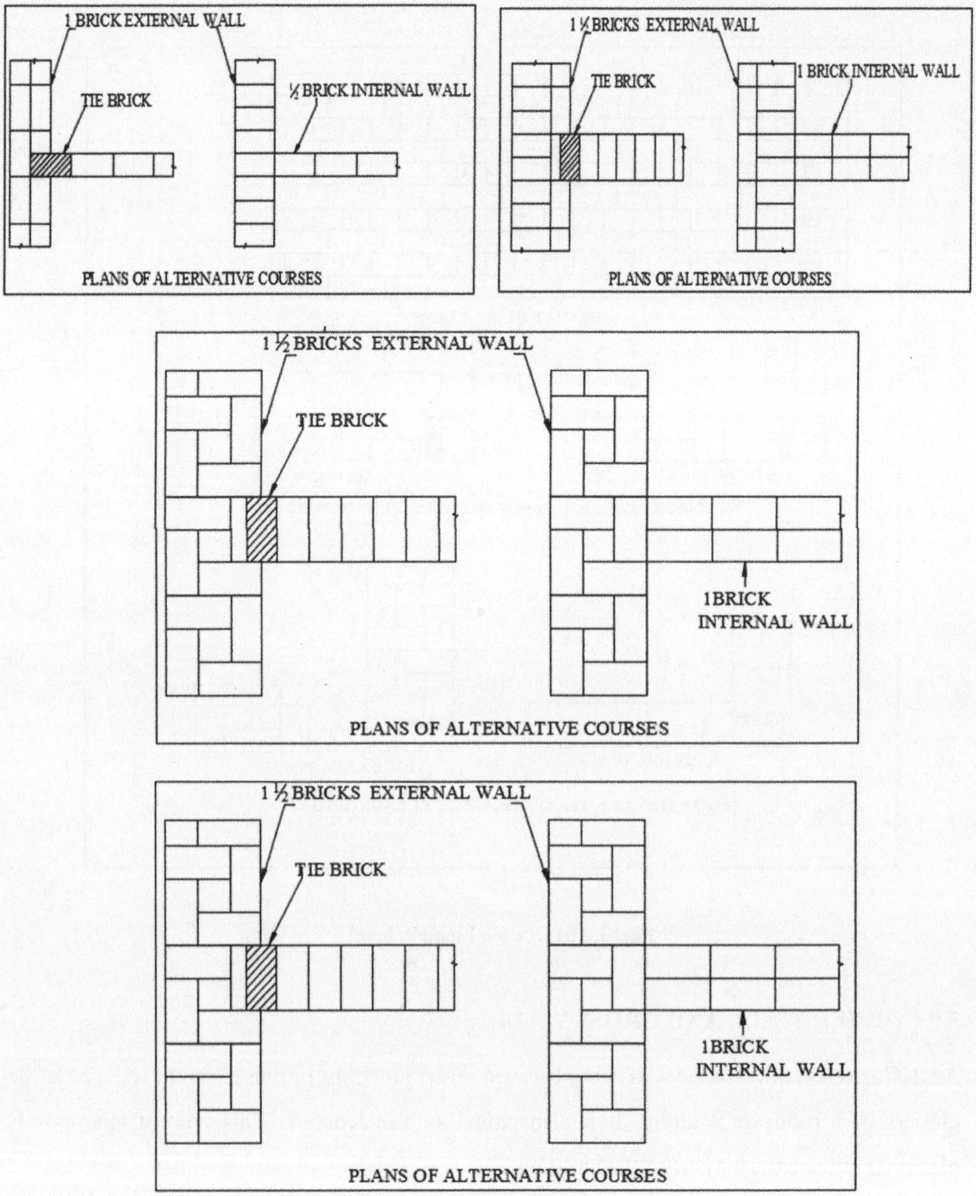

Fig. 3. 16 : Corner Wall

3.5.1 Cross Wall: Cross walls are observed at the junctions of two different walls. It is also called as cross junction. This type of junctions is observed in the interior part of the building where both the walls are continuous.

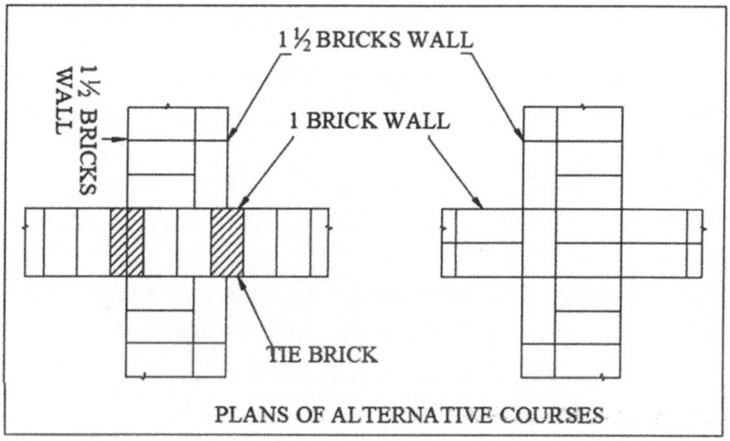

(a)

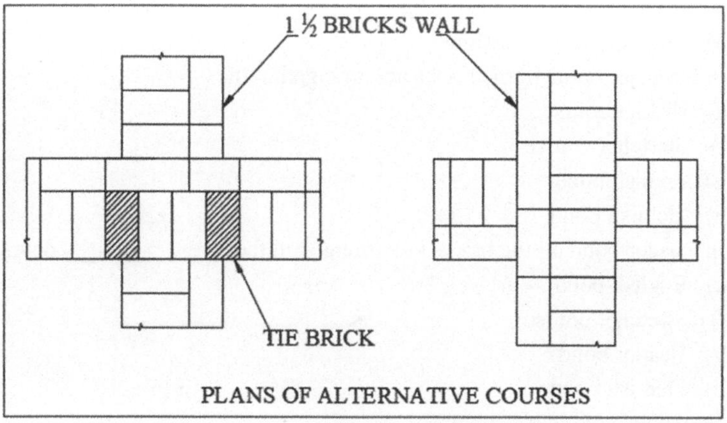

(b)

Fig. 3. 17: (a) and (b) Cross wall

UNIT SUMMARY

This module makes the student learn about various masonry bonds used in the construction of walls with a variety of thickness. The salient features to remember are

- The technical terms and various faces of the blocks used
- The most important is the bond pattern and block-laying quality to act on the masonry unit as one.
- The types of bonds to be used for different position of walls
- Basics of good design principles for effective load-carrying capacity & architectural views.
- The importance of through stones at the junction of cross walls and corner walls.
- Effective bonding of all the walls with different thicknesses at the junction.

EXERCISES

A. Multiple Choice Questions

1. In Stretcher bond all the bricks are arranged in the _____________ courses
   e) Header course
   f) Stretcher course
   g) English bond
   h) Flemish bond
2. In Header bond all the bricks are arranged in the _____________ courses
   e) English bond
   f) Stretcher course
   g) Header bond
   h) Flemish bond
3. Strongest bond in brick work_____________
   e) English bond
   f) Stretcher course
   g) Header bond
   h) Flemish bond
4. The number of mortar joints in the header course of English bond is _____________
   e) One time of stretcher course
   f) Double of stretcher course
   g) One and half time of stretcher course
   h) Three times of stretcher course
5. The Face elevation of single Flemish bond is of_____________
   e) English bond

   f) Flemish bond
   g) Stretcher bond
   h) Header bond
6. The back of single Flemish bond is of____________
   e) English bond
   f) Flemish bond
   g) Stretcher bond
   h) Header bond
7. The minimum thickness for single Flemish bond is ____________
   e) 1 Brick wall
   f) 1 ½ Brick wall
   g) 2 Brick wall
   h) 2 ½ Brick wall
8. The Face and back elevation of double Flemish bond is of____________
   e) Flemish bond
   f) English bond
   g) Stretcher bond
   h) Header bond

Answers

1. b 2. c 3. a 4. a 5. b 6. a 7. b 8. a

B. Short and Long Answer Type Questions

3.1 List out the advantages and disadvantages of English bond and Flemish bond?

3.2 Illustrate with neat sketch a) stretcher b) Header c) Closer d) Bat e) Frog

3.3 Differentiate between English bond and Flemish bond?

3.4 Write a short note on different types of brick masonry?

KNOW MORE

- *Rat Trap Bond:* Bricks are laid on the edge of the wall, creating a gap in the thickness of the wall, which saves 25% of the bricks compared to a solid wall.
- *Composite masonry:* walls are built of two different materials like stone cum brick, brick cum SMB, etc. The main purpose of this masonry is to reduce the cost and also the thickness of walls. Generally moderate materials are used inside of the wall, and highly durable materials are used on the outer face of the wall to effectively resist the environmental impacts.
- *Cavity walls:* These walls are commonly used for thermal insulation. These walls may go beyond 40 cm thickness based on the exposure conditions and thermal insulation required. These walls are mainly constructed on the faces of buildings exposed to the sun.
- *Rammed Earth Technique:* The construction of in-situ walls using_the plain earth (or stabilized earth) is named as rammed earth walls. These walls can be specifically used for buildings of one storey. Unreinforced rammed earth construction is gaining more recognition in eco- construction of buildings.

REFERENCES AND SUGGESTED READINGS

1. Rangwala. Building Construction, 34th ed.; Charotar Publications, 2022.
2. K S Jagadish, B V Venkatarama Reddy and K S Nanjunda Rao. Alternative Building Materials and Technologies, 3rd ed.; New Age International Publishers, 2024.
3. K S Jagadish. Structural Masonry, I K International Publishing House Pvt. Ltd., 2015.

Dynamic QR Code for Further Reading

4 BUILDING DRAWING

UNIT SPECIFICS

The following aspects are discussed in this module.

- Different terms used in building drawing
- How to start creating a plan from the beginning to the client's ultimate imagined specifications.
- Showing various sections and elevations of the plan
- Specifying the materials to be utilized for each element.

RATIONALE

This chapter explains to students the basic sizes required for each element of the building and how to plan a building so that they can perfectly convey their thoughts and vision with all of the available tools and materials.

PRE-REQUISITES

- Elements of a building
- Construction materials and techniques
- Bye laws
- Basics of Vaastu

UNIT OUTCOMES

The following are the outcomes of this unit:

U4-O1: Understand the space requirements of different rooms.

U4-O2: Plan a building with all the requirements in proper orientation.

U4-O3: Show the elevation and section of the building.

U4-O4: Specify different materials to be used for effective construction.

Unit- 4 Outcomes	EXPECTED MAPPING WITH COURSE OUTCOMES (1-Weak correlation; 2-Medium correlation; 3-Strong correlation)				
	CO-1	CO-2	CO-3	CO-4	CO-5
U4-O1	1	1	3	2	3
U4-O2	2	2	3	3	3
U4-O3	2	2	3	3	3
U4-O4	2	2	3	2	2

Building drawing is the skill of utilizing available space by properly organizing various rooms. As a result, tenant or occupant of the building feels pleasing in utilizing the area provided. The art of building drawing enables the architect to present his or her vision of proportioning and placing various units of a building with ventilation and effective use of natural light. This chapter covers all of the principles and procedures for preparing a building drawing.

4.1 TERMS

In building drawing various terms are used to describe different areas and ratios. The following are definitions of some common terms

(i) **Carpet Area:** The real usable space inside an apartment or building walls apart from the thickness of the inner walls is referred to as the carpet area. It excludes common areas such as the lobby and staircase and includes the net usable area inside the walls.

(ii) **Plinth Area:** The built-up covered area measured at a building ground level is called the plinth area. It is the entire ground floor area of the building, including the thickness of the external walls. This space is measured from the outside face of the walls up to the floor slab level.

(iii) **Built-up Area:** The built-up area includes the carpet surface as well as the space occupied by walls, balconies and other structures.

(iv) **Super Built-up Area:** It includes the built-up area of a multi-unit complex as well as common areas like the lobby, elevator, staircases and other shared facilities. It is commonly used to determine the estimate of a structure.

(v) **Total Floor Area:** It is the total floor area of all the floors in a structure. It can be estimated by combining the individual floor areas, including all levels such as basement, ground floor, first floor, second floor etc.

(vi) **Gross Floor Area:** The entire floor area inside a building, including all levels and spaces is referred to as the gross floor area. It is usually measured from the outside faces of the walls. It includes the ground level and upper floors as well as all functional spaces inside a building such as intermediate floor (mezzanine floor), attic floor, rooms, and basements.

(vii) **Floor Area Ratio (FAR):** The floor area ratio indicates the relationship between the total floor area of a building and the total area of the land or plot on which the building standards. It is also called as Floor Space Index (FSI).

Floor Area Ratio = Total Floor Area/ Total Land Area

(viii) **Plot Ratio:** The plot ratio is the relationship between the total floor area of a building and the total area of a plot of land. It is similar to the floor area ratio. However, depending on the local restrictions, certain sorts of areas may be excluded such as basements.

(ix) **Coverage Area Ratio:** It is the ratio between the building's ground coverage area and the total area of the land.

Coverage Area Ratio = Ground Coverage Area/ Total Land Area

(x) **Open Space Ratio:** It is the ratio between the built-up area to the open space on a plot. It represents the amount of open or green space available relative to the developed area.

Open Space Ratio = Open Space Area/ Built-up Area

4.2 ELEMENTS OF PLANNING BUILDING DRAWING

The various elements of planning building drawing are Structural elements and Non-structural elements

4.2.1 Structural Elements

(i) **Foundations:** The foundation is the main part of a structure that transfers the load from super structure to the soil beneath, uniformly based on the safe bearing capacity of the soil. These are different types of foundation used

(a) Shallow Foundation: The depth of footing is lesser (or) equal to the width of footing. Spread footing, isolated footing and combined footings are the examples of shallow foundation.

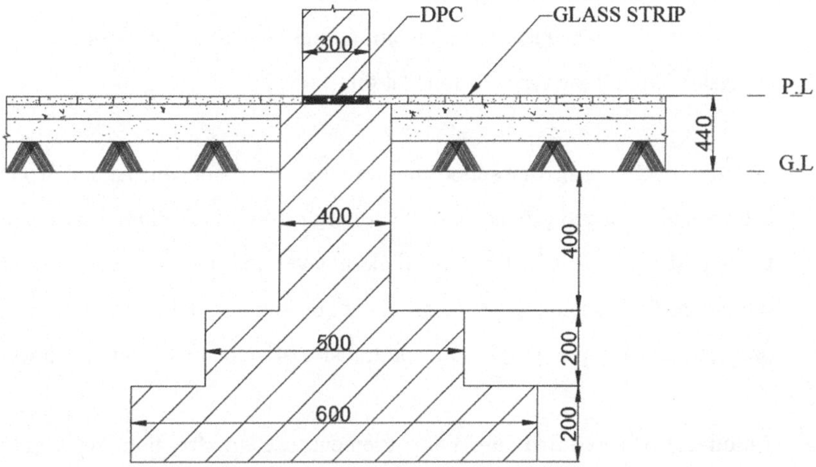

Fig. 4. 1: Spread Footing

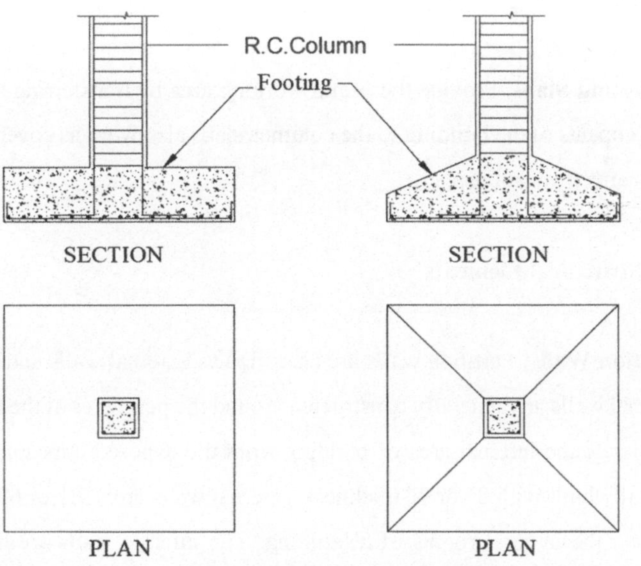

Fig. 4. 2: Isolated Footing

(b) **Deep Foundation:** If depth of footing is greater than 1.5 times width of the footing. It is called as deep foundation. Pile foundation and well foundation are classified in these types of foundation

(ii) **Plinth:** An RCC beam or a supporting wall used to separate the natural ground level and ground level and ground floor level is called plinth. A Plinth beam is essential to support wall loads at the ground floor level and also to reduce the effect of settlement from ground level to ground floor level. A damp-proof course is also provided on the top of plinth, to avoid transfer of dampness from soil to walls.

(iii) **Columns (or) Load Bearing Walls:** Elements used to carry floor loads and transfer them safely to the foundations. These are the more essential and load carrying elements necessary in the construction of a building. The position and stiffness of these members play an important role in the load resisting performance of a building.

(iv) **Beams and Slabs**: Provide the useful working area by transferring all the loads of the occupants of the building to the columns slabs also work as coverings to protect the tenants from atmosphere.

4.2.2 Non-Structural Elements

(i) **Partition Walls:** Partition walls are classified as external walls and internal walls. External walls are generally constructed around the perimeter of the building frame to separate the internal area of building from the exposed atmosphere. These are generally built with 8" or 9" thickness. Internal walls are 41/2" or 6" thick, built to separate the internal rooms of a building. The internal walls are not exposed to atmospheric conditions.

(ii) **Doors and windows:** Doors provide access to different rooms of the building and windows are provided to give natural ventilation inside the building.

a) Terms used in doors and windows

Frame: It is made up of many wooden or steel elements that support a door or window. Door frames are made up of one horizontal and two vertical members, whereas window frames are made up of two horizontal and two vertical parts.

Jamb: It is a vertical face of a door or windows opening which supports the frame.

Sill: It is the bottom most horizontal member forming the frame.

Top rail: It is the uppermost horizontal member of the shutter.

Bottom rail: It is the lowermost horizontal member of the shutter.

Lock rail: It is the central horizontal member of the shutter.

Intermediate rail: It divides the door or window frame into number of panels.

Frieze rail: It is the horizontal member between top rail and lock rail of the shutter.

Horn: It is a horizontal projection of head member or sill beyond the face of the door or window frame.

Style: It is an external vertical member of the shutter.

Mullion: It is the vertical member used to increase the panels running through a shutter.

Holdfast: It is made of steel and provided on both sides of the window frames or vertical door to keep the frame in position.

Louver: It is a timber piece, which is attached in inclined position within frame.

Molding: The wooden pieces are used to fill the space between the two vertical members and two horizontal members.

Sash: It is a light section frame that is used for fastening the glass in the panels. It is made up of two horizontal and two vertical members.

The general sizes of the doors for residential buildings are:

Main door	1200mm x 2100mm
Side door	1000mm x 2100mm
Kitchen door	900mm x 2100mm
Toilet door	750mm x 2100mm
Garage door	2500mm x 2300mm

b) Types of Doors

Battened and ledged door: These doors are built of timber known as batten. The dimensions of the battens are 150mm to 300mm wide and 25mm thick, are positioned vertically side by side and are held in place by horizontal member known as ledges. The width and thickness of the edges ranges from 100 to 150mm.These doors are quite basic and used in small spaces like restrooms.

Framed and paneled door: It is made with a framework that has panels installed in it. From top to bottom, there is continuity in the styles. The style is connected to the top, middle, and bottom rails. Additionally, mullions are offered to minimize panel size. These doors are standard and found in all structures. They also improve the door's appearance.

Glazed and panelled door: These kinds of doors are available when additional light than that offered by the windows is required. Using putty and tiny nails, the glass is fastened into the rebates set up in the wooden sash bars. These doors are generally provided in classrooms, offices and hospitals among other places.

Flush door: Flush doors are provided for smooth and clean surface. These doors are very strong and durable. It consists of a frame work of timber styles and rails with plywood covering both the faces of the frames. There are three types of flush doors: laminated, hollow and framed doors.

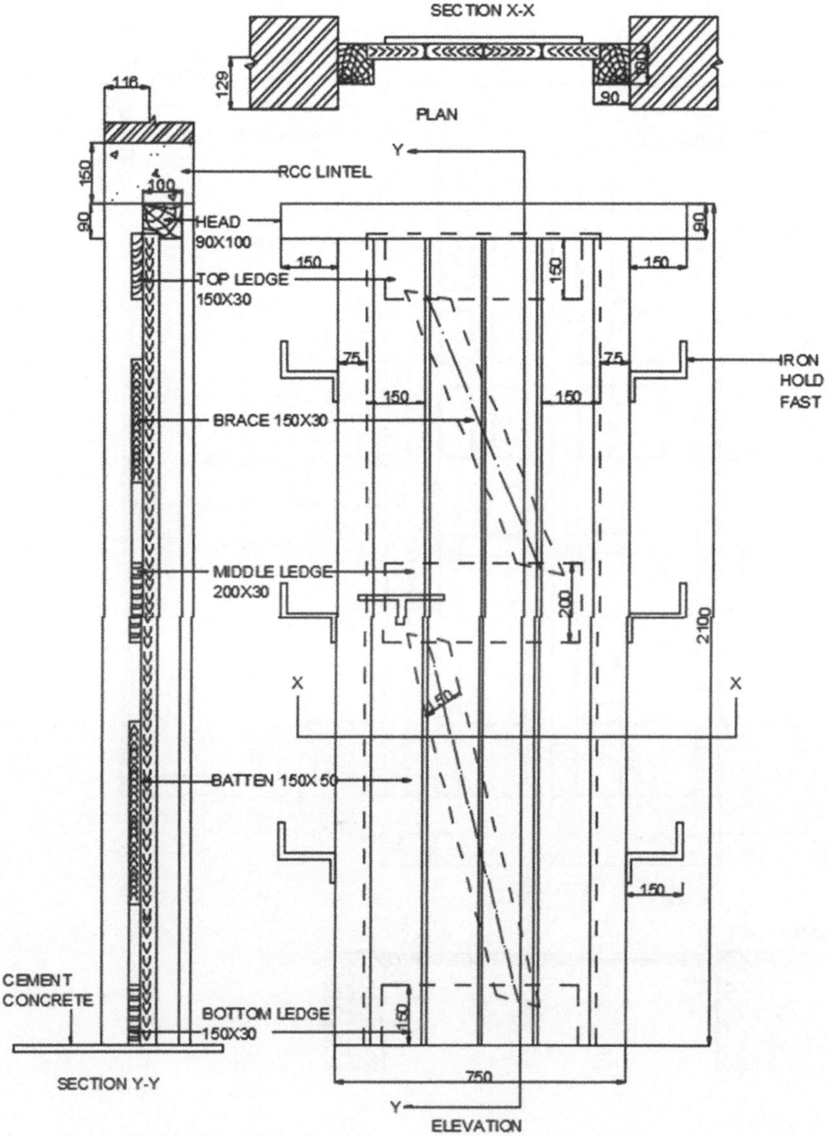

Fig. 4. 3: Battened ledged and braced door

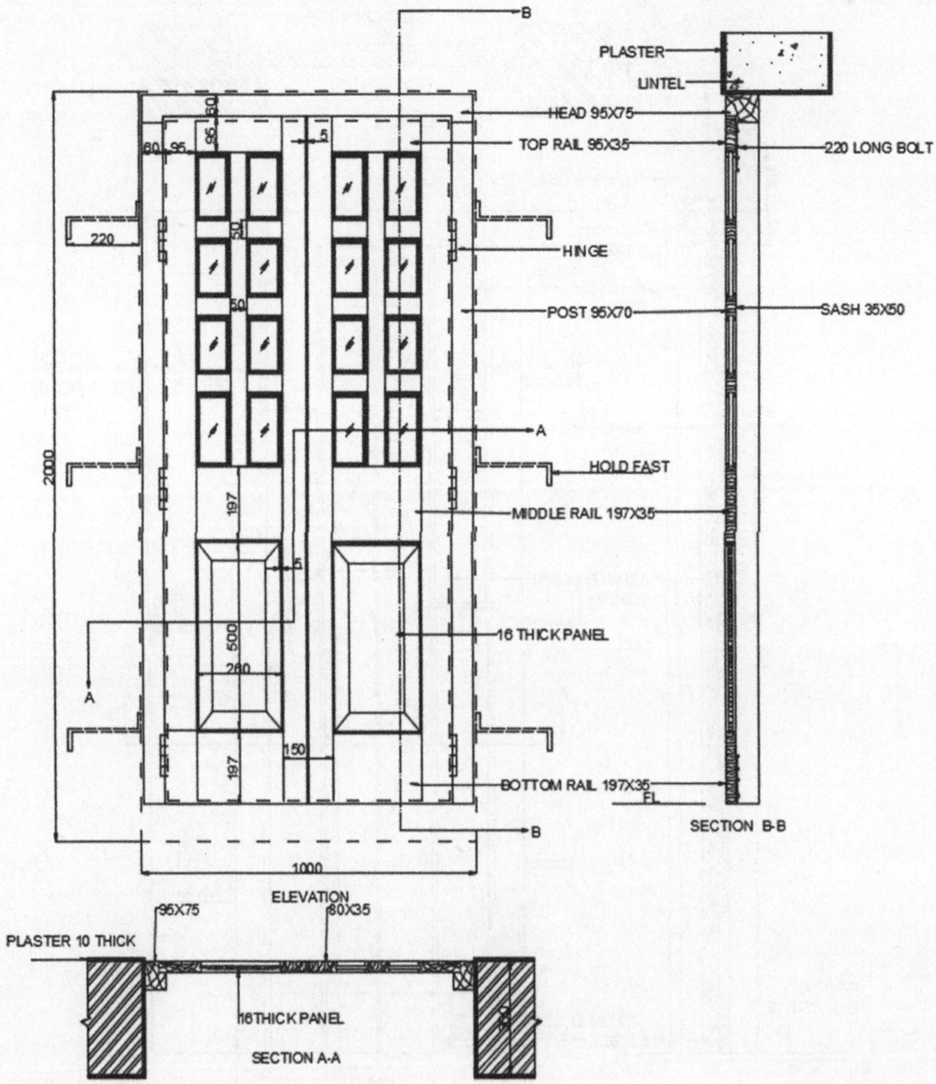

Fig. 4. 4: Glazed and panelled door

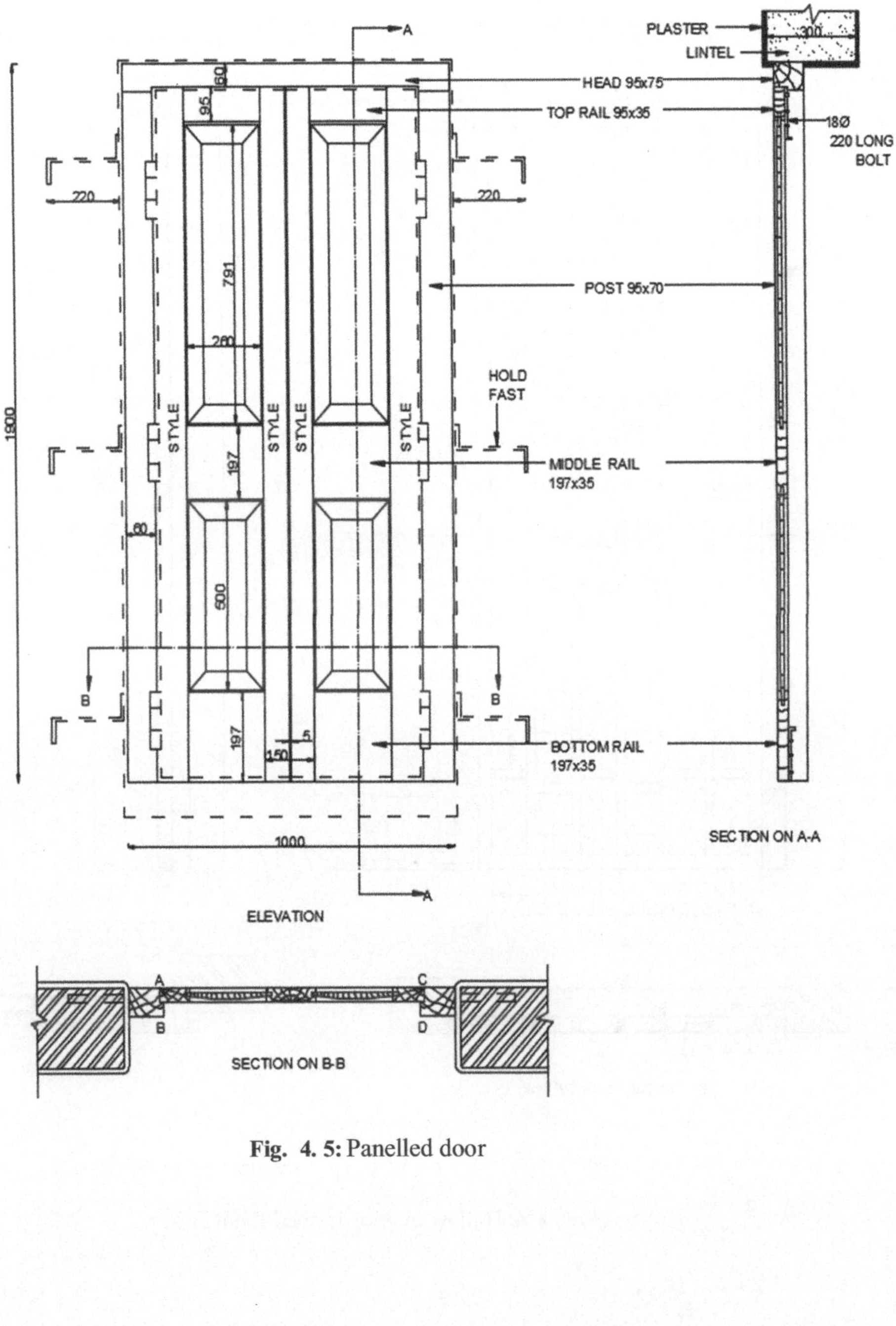

Fig. 4. 5: Panelled door

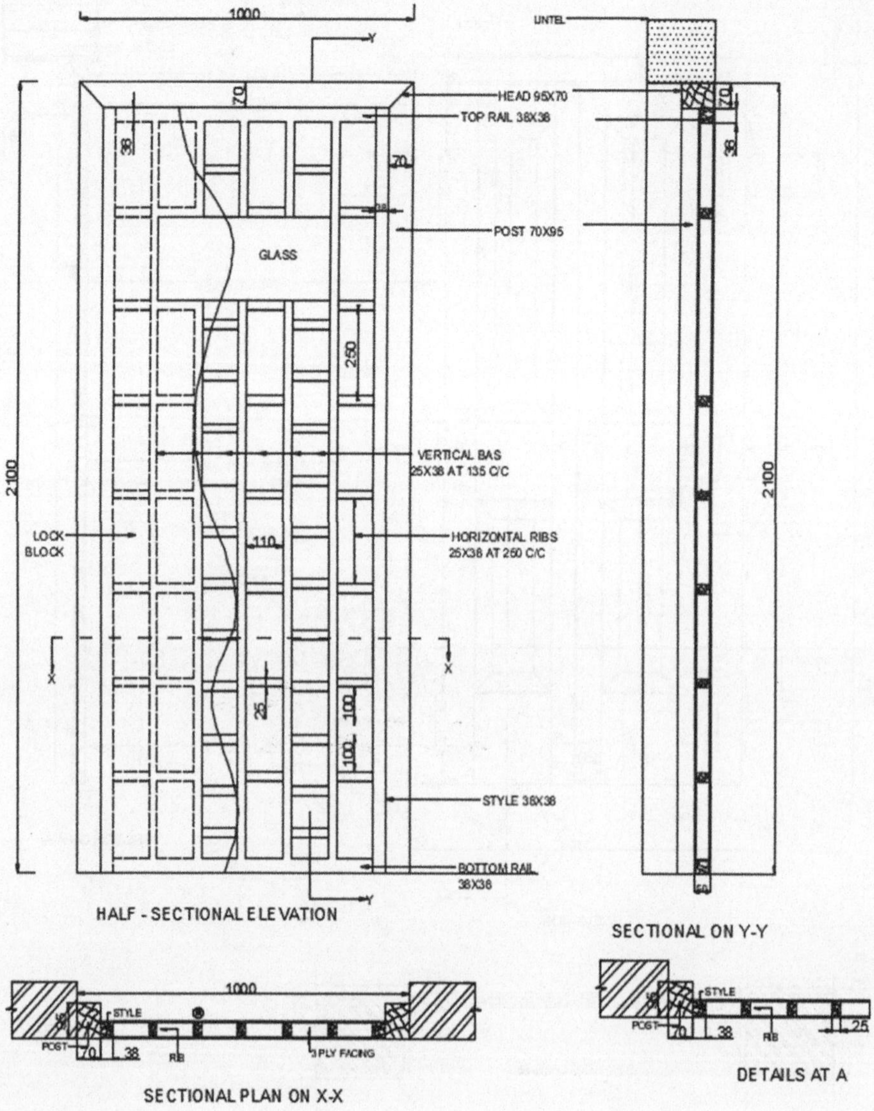

Fig. 4. 6: Hollow core or framed flush door

c) Types of windows

Windows are provided in the building to allow in natural light and fresh air. In residential buildings, the window's area should not be less than 1/10 of floor area and for public building it may be taken as 1/5 of floor area. A ventilator is also a small window that is positioned above the lintel of door or window to facilitate the discharge of hot air.

Panelled window: The construction of panelled window is same as the panelled door. The size of style and horizontal rails are 45 to 100mm wide and 25mm thick in section. The window shutter is made of wooden panels or ply wood panels.

Glazed or Sash window: It is similar to panelled window except that glass panels are fixed in sash bar with the help of putty or nails.

(iii) **Lintels, sills and Chejja:** Doors and windows should never be placed directly on the masonry, a 75mm thick plain concrete bed is to be provided over the masonry as a supporting layer called as sill.

Lintels are beams used to transfer the load over the doors/windows safely so that there is no effect of walls above on the doors. Lintels can be RCC beams or simple stone beams.

Chejjas is the projection given outside the wall to project doors and windows from the rain.

(iv) **Stairs:** Stairs are used to provide access vertically in a building (floor to floor). There are various types of staircases provided based on the space and access required in the building.

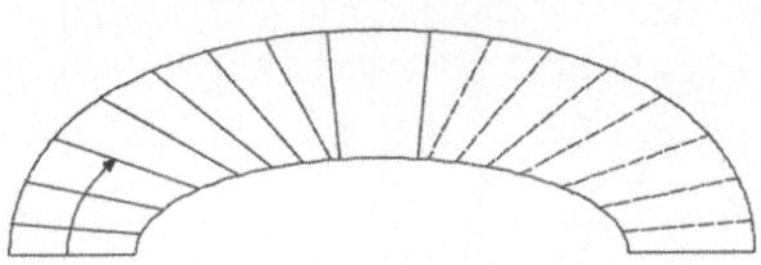

ELLIPTICAL

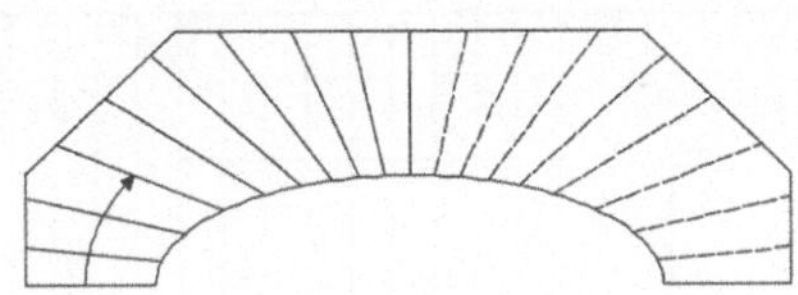

SEMI ELLIPTICAL

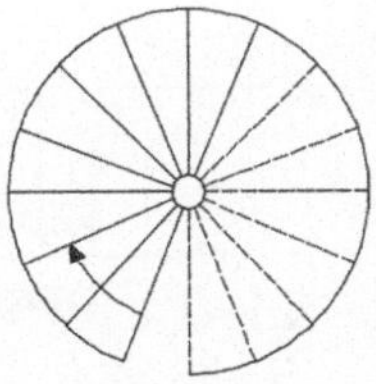

SPIRAL

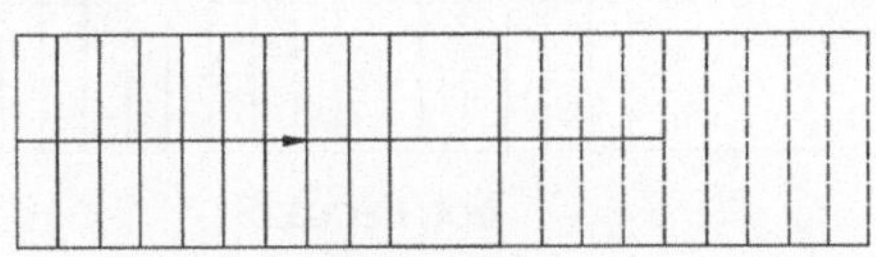

STRAIGHT FLIGHT

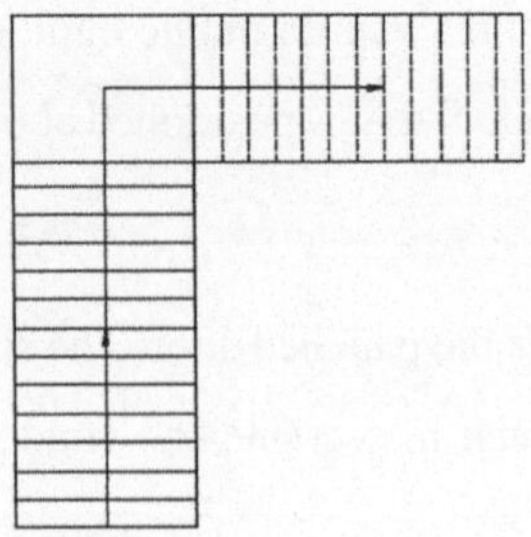

QUARTER TURN

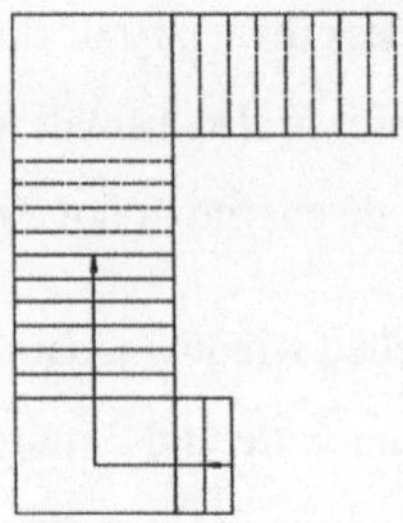

QUARTER TURN

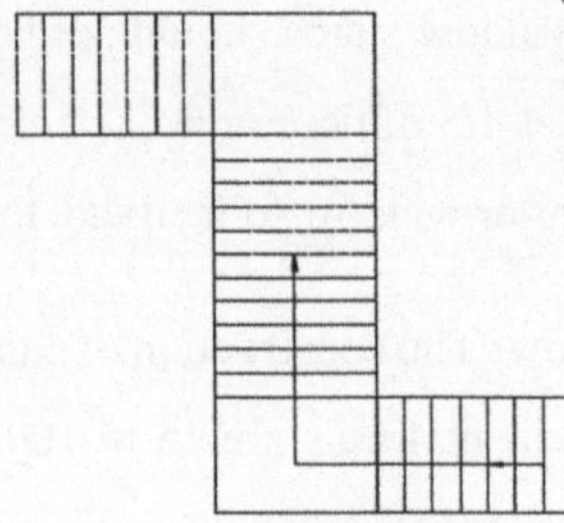

QUARTER TURN

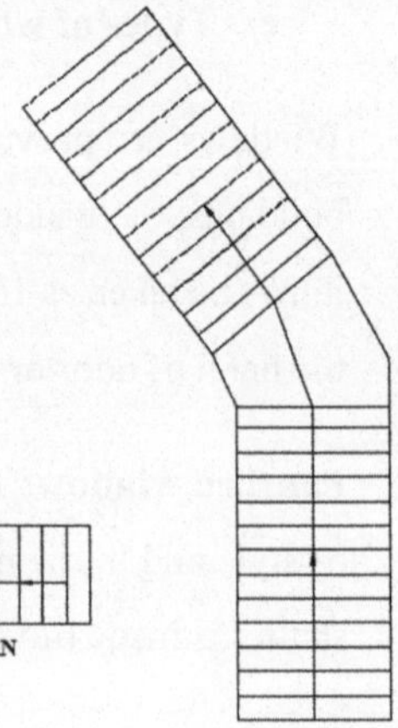

QUARTER TURN

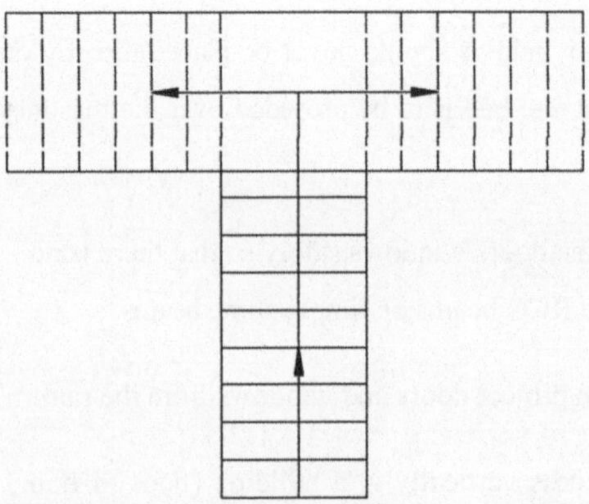

BIFURCATED

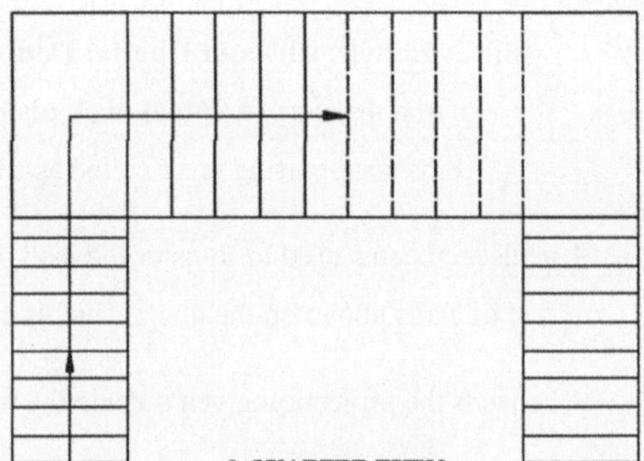

3-QUARTER TURN

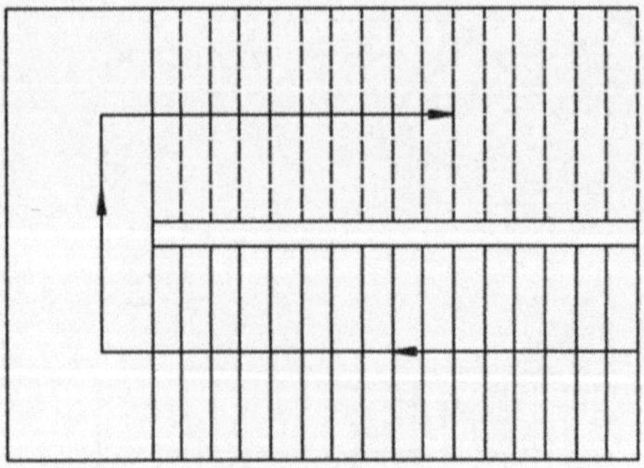

DOG LEGGED

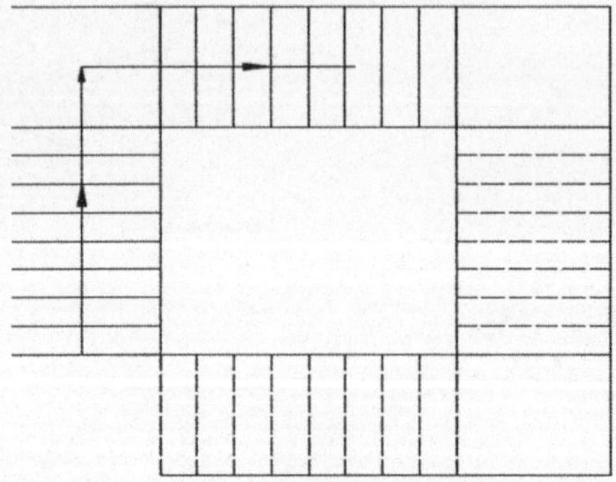

OPEN WELL

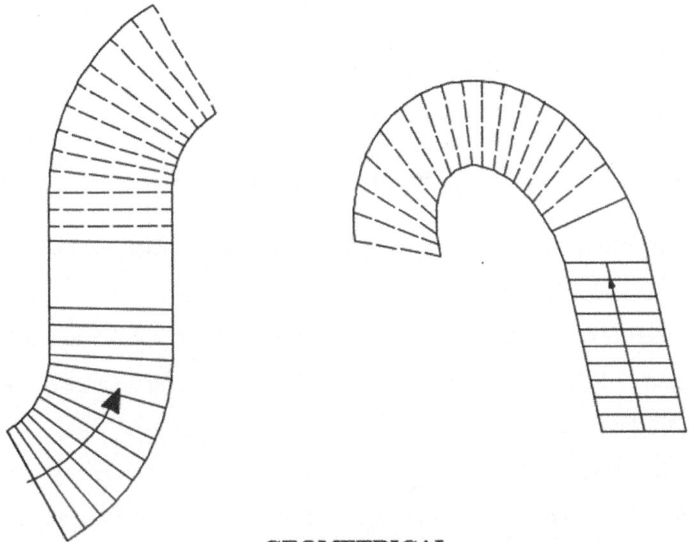

GEOMETRICAL

Fig. 4. 7: Stairs

(v) **Floor Finishes/ Flooring:** The process of covering over a floor to make the space more aesthetical is known as floor finishing. There are vast varieties of flooring materials used, some of them are listed below

(a) **Mud Flooring:** Usually used in low-income group houses. It is very hard, cheap, and fairly impervious.

(b) **Brick Flooring:** It is cheap, durable, and easily repairable. Used in warehouses, godowns where heavy articles are stored.

(c) **Mosaic Flooring:** This flooring is decorative and damp resistant. Used in temples, costly houses, and monumental structures.

(d) **Tile Flooring:** It is very common these days. User friendly and damp proof. It can be used in any type of building.

(e) **Marble Flooring:** Decorative and durable. Easy to maintain. Used in residences, hospitals, public building and temples for architectural appearance.

(f) **Glass Flooring:** It is both costly and attractive. Used for the best illumination in showrooms.

(g) **P.V.C Flooring:** It is hard, attractive, and resilient. Used in hotels, hospitals and shops.

(vi) Roofs:

The upper part of the building is used to cover the area and protect it from natural elements like sun, wind, snow, rain, etc.

The different varieties of roofs are presented below.

(a) Flat roof: A flat roof is suitable for multistorey structures and in places of hot regions and minimum rainfall.

(b) Pitched (or) sloping roofs: Used in places where rainfall, snowfall is heavy. It is considered a low-cost routing when spans are high and no access is required to the top of the floor.

(c) Domes (or) Shells (or) Covered roofs: Used the majority for architectural appearances. Very efficient in large circular walls used for assemblies, gymnasium etc.

4.3 METHODS TO DRAW THE LINE PLAN

The first stage of preparing a drawing for a building is preparing a line plan, which shows whether all the functional requirements of the building are placed and spaced in a proper way. The line plan represents the center-to-center dimensions of the proposed drawing. Before preparing a line plan, the plinth area of the building has to be determined based on the setbacks given by the local authorities of the city.

The sets involved in drawing a line diagram are

i) List out all the functionalities of the building and their standard space requirements, along with their appropriate location, as given in Table 4.1

Table 4.1: Size requirements and positions of various rooms in a building

Type of Room	Size Required	Suggestible Position as Per Vaastu
Living Room	4.2m X 4.2m	North-East/ North/ East
Bed Room	3.6m x 3.6m	Master Bed Room – South-West Children Bed Room – North-West
Dining Room	3.6m x 3.8m	South/West
Kitchen	2.7m x 3m	South-East
Bath Room	1.2m x 1.8m	South/West

Guest Room	3m x 3m	North/West
Store Room	2.5m x 2.5m	South/West
Pooja Room	1.5m x 1.8m	North-East

ii) Resemble the directions of the building (North), road way, and width to the plot.

iii) Place all the required rooms according to the space required, giving due consideration to aspect ratios, ventilation, convenience to access the proposed room, comfort to the tenants, and following the principles of planning such that the building is aesthetically strong.

iv) For the obtained line diagram, place all the doors, windows, ventilators, and staircases with standard dimensions as shown in figure 4.8.

v) Specify all the sizes of the rooms assumed and the sizes of the elements assumed as references. The dimensions represented in the line plan are center-to-center dimensions and should be clearly marked as "c/c" at the end of the dimension.

vi) The scale of the dimensions has to be clearly mentioned on the line diagram drawn.

Note: A dimension of the room mentioned, DINING ROOM, is 3m x 4m c/c, which indicates 3m as a horizontal dimension and 4m as a vertical dimension.

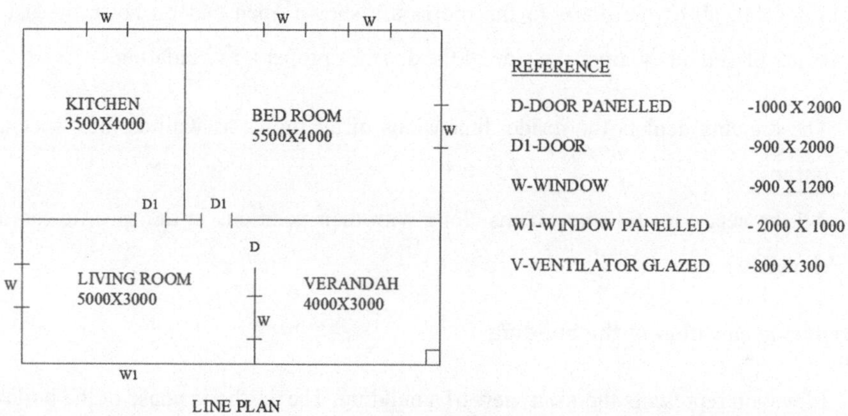

Fig. 4.8: Line plan

4.5 DETAILED DRAWING

After the line plan is finalized, a detailed plan with elevation and different sections must be developed to ground the project. The steps to be followed are

(a) Preparing a detailed plan

- The thickness of all the walls has to be fixed based on the location and exposure conditions.

Note: These are two types of walls as mentioned earlier, external walls (9" or 8" thick) and internal walls (4.5" or 6" thick)

- Keep outer lines (boundary lines) of the building unchanged, the thicknesses of all the external walls are added.
- The internal walls are drawn in two different ways, the first approach involves drawing lines on both sides of the internal wall in a line plan with equal distance, and second method involves drawing a line with an assumed thickness on the inside of the room.
- After all the walls are drawn, place doors and windows to allow access to other spaces and ensure sufficient ventilation.

Note: In general, all of the doors to the rooms are shown open and on north or east side. Windows are placed on the opposite side of the door for proper air circulation.

- The drawing depicts the inside dimensions of all rooms as well as their specialized purpose.
- All the necessary finishing items along with their positions in the specific rooms are shown.

(b) Preparing elevation of the building

Elevation represents the front view of a building. The aesthetic phase of the building is clearly visible in elevation. Many of the clients look for good elevation for the building. Preparing elevation depends on purely the vision of the architect working on the building with due consideration to client's requirement. Some considerations/points to be keep in mind are listed below

- The elevation of a building should never disturb the orientation of the rooms.

- The elements/materials used in the elevation should be easily available and easy to maintain.
- The structural engineer must ensure the stability of projections or extensions of structural elements such as cantilever slabs and porticos before finalizing.
- A simple and pleasing elevation without disturbing the plan of the building is always recommended.

(c) Preparing sections of the building

A section represents the longitudinal and lateral part of the building in vertical plane. A section is drawn to specify the vertical dimensions and positions of various elements in the building.

- The section is to be drawn along the length or width of the building and should show all the specifications of the proposed building from sub-structure to super-structure both structurally and non-structurally.
- A clear drawing of all the important elements, including with the sizes and exact position along the height is to be shown.
- The section made should show all of the floor level positions as well as all of the floor level positions with respect to the (NGL) natural ground level. Hence a proper drainage plan can be presented depending upon the intended floor levels.

4.5 SITE PLAN, FLOOR PLAN, ELEVATION AND SECTION DRAWING OF SMALL RESIDENTIAL BUILDINGS

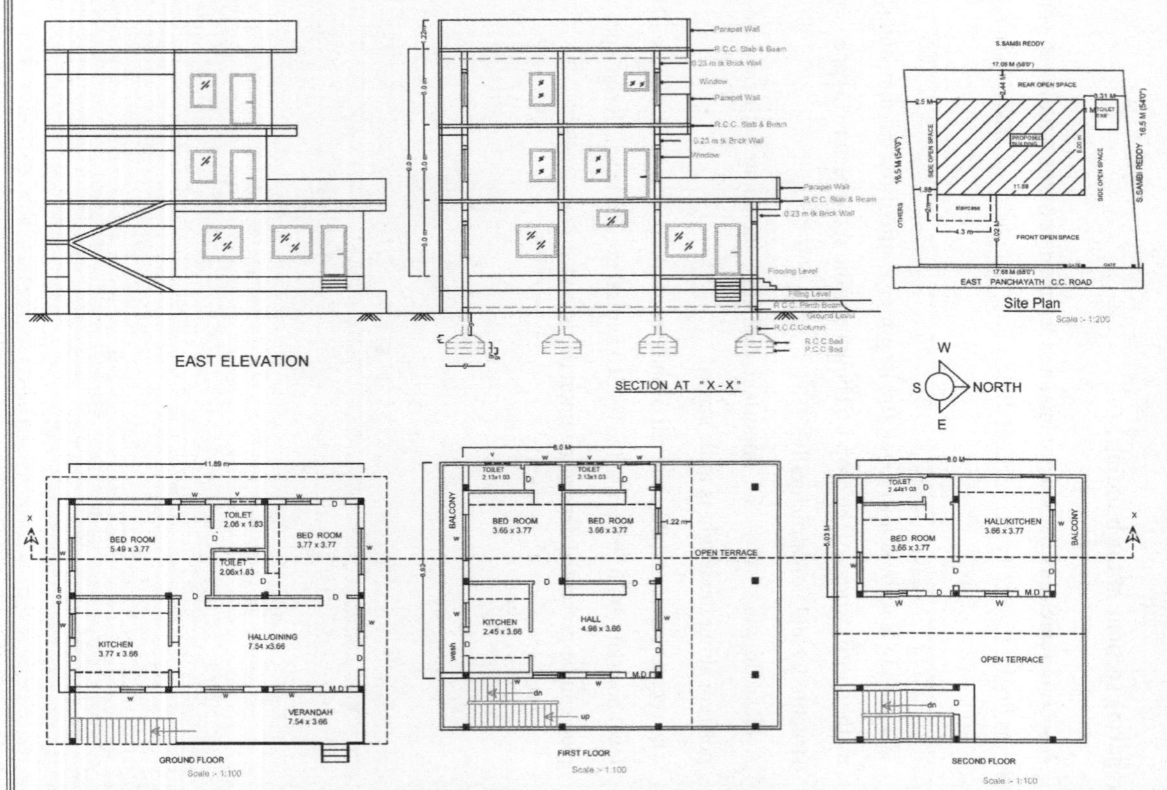

OFFICE USE:

Floor Space Index :-
Site Area = 290.96 Sq.mt(348 sq.yds)
Plinth Area = 95.12 ,,
Toilet Area = 4.45 ,,
Open Area = 191.39 ,,

Areas:-

Ground Floor Plinth Area	= 95.12	Sq.m
First Floor Plinth Area	= 71.36	''
Second Floor Plinth Area	= 40.24	''
Total	=206.72	Sq.m

Joinery:-

Doors:-	Windows:-
M/D = 1.06 x 2.06	W = 2.13 x 1.45
D = 0.91 x 2.06	W1 = 1.06 x 1.45
D/w = 2.13 x 2.06	Wk = 1.06 x 0.76
D1 = 0.76 x 2.06	V = 0.61 x 0.91

Note:-
All dimensions are in Meters
Scale 1:100 & 1: 200
Pro. work shown Thus = ▭

Applicant :-

Architect :-

TEMPLATE NO 1

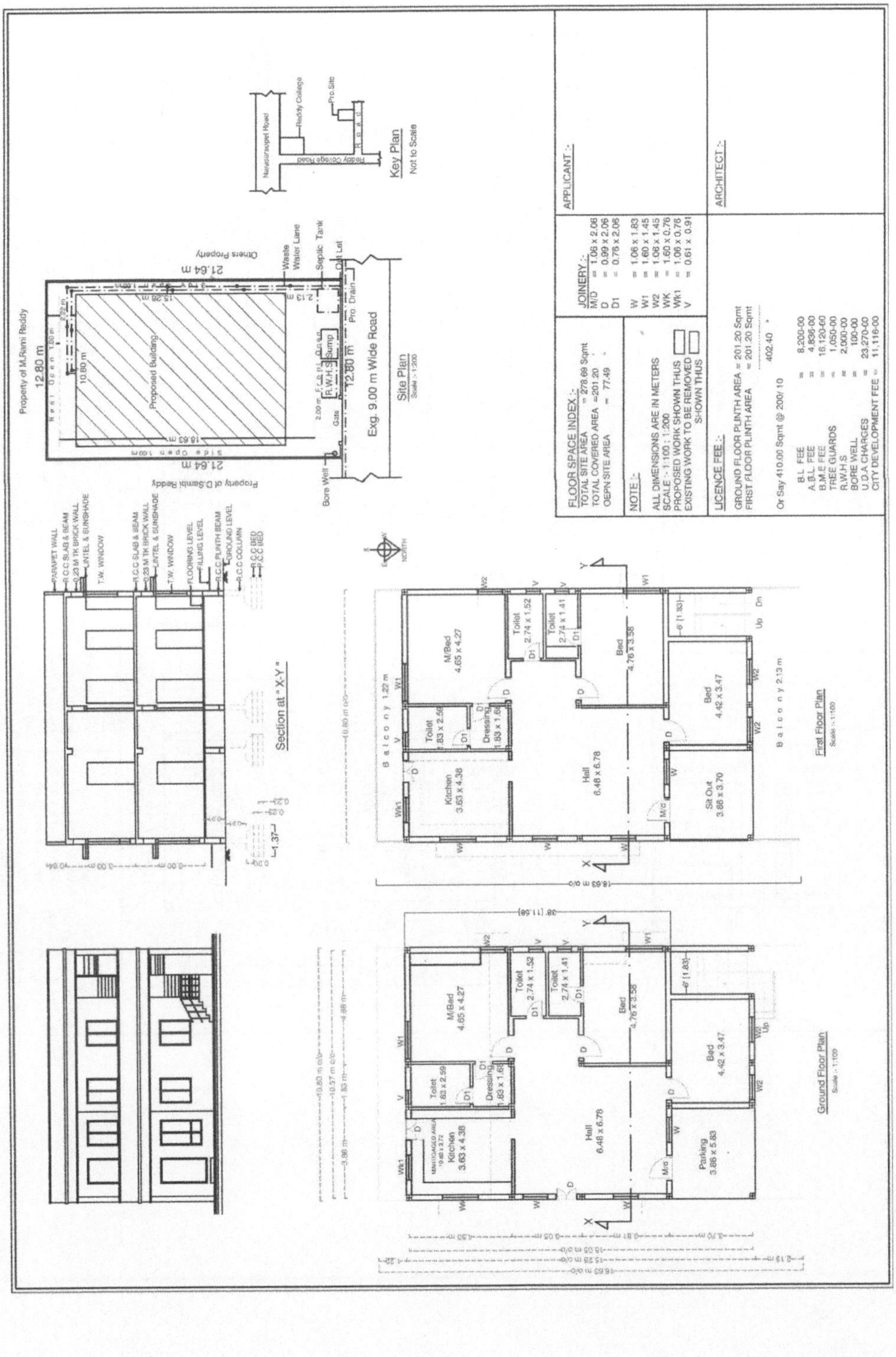

TEMPLATE NO 2

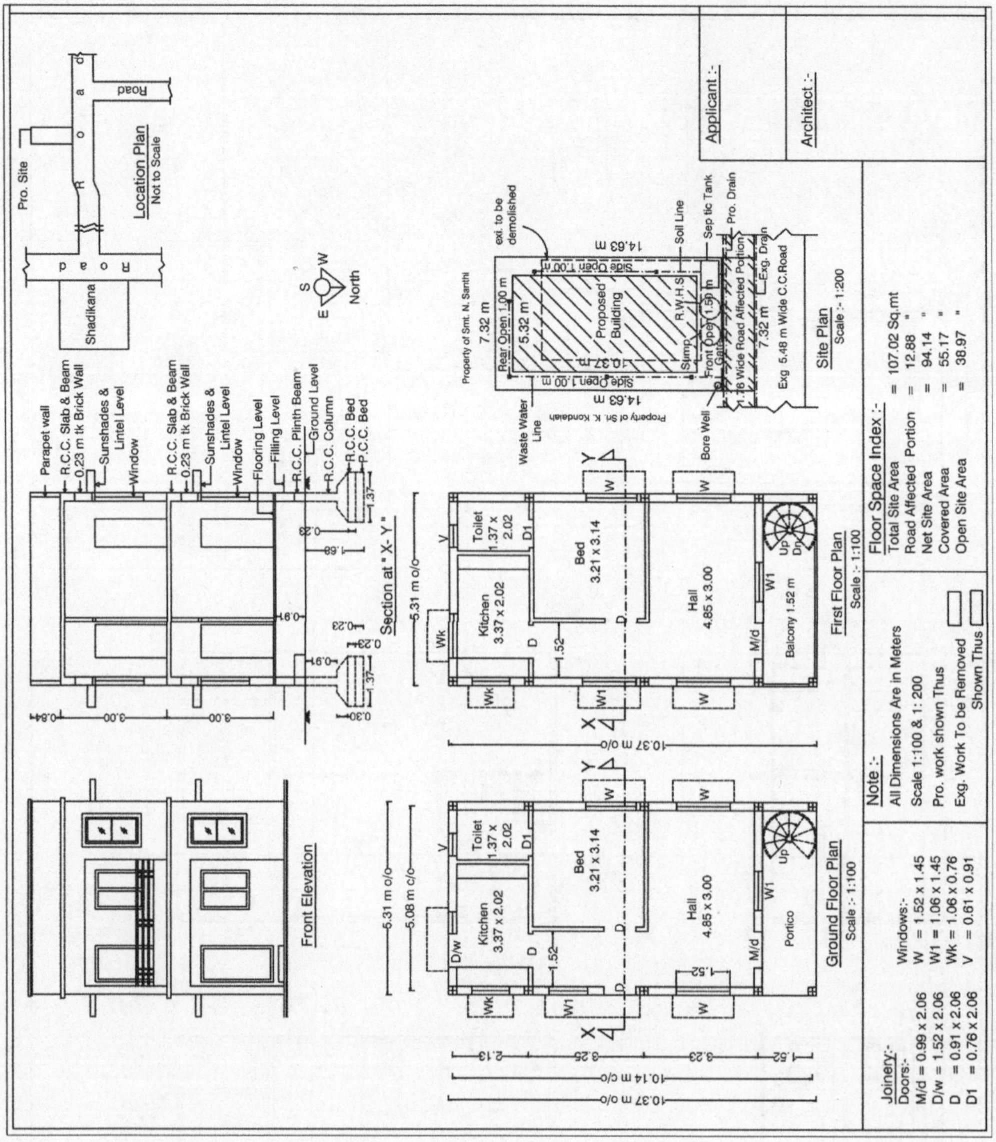

TEMPLATE NO 3

4.6 FOUNDATION PLAN

The footing plan indicates the size of the footing to be provided for each column. The center of the column should match with the centre of footing. An example of showing footing plan is given in figure 4.9.

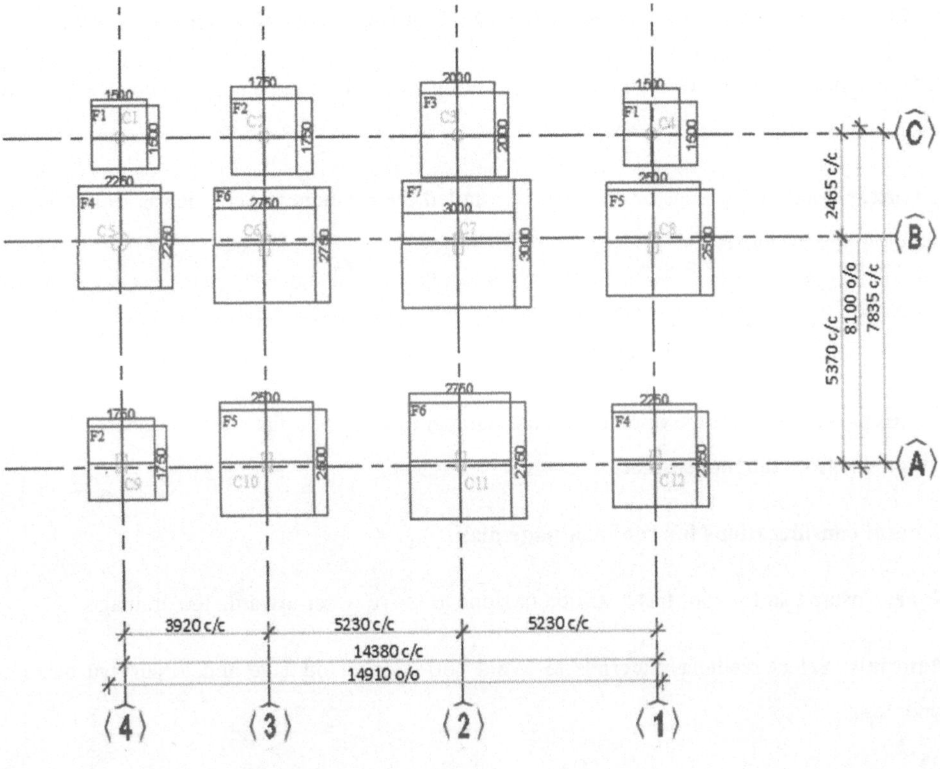

Fig. 4.9: Footing plan

4.7 ROOF DRAINAGE PLAN

A roof drainage plan is used for the effective disposal of rainwater away from the building without affecting any components of the building. The main roof drainage components are shown below.

Key components of the roof drainage plan

1. Gutters: To collect and direct rainfall, channels are built along the edges of the roof.

2. Downspouts: These are the pipelines that bring water from the gutters down to the ground level.

3. Gutter Guards: Screens or covers are installed over gutters to keep debris from entering and clogging the system.

4. Splash Blocks: Devices are installed at the bottom of the downspouts to direct water away from the foundation.

5. Scuppers: These are the openings in the parapet or edge of a flat roof that enable water to drain away from the roof surface.

General considerations for roof drainage plan

Slope: Ensure that the roof has a sufficient slope to move water towards the drainage.

Materials: Select resilient materials to avoid corrosion or damage and to survive weather conditions.

Capacity: Design gutters and downspouts to handle anticipated rainfall volume.

Maintenance: Plan for frequent inspections and cleanings to prevent blockages and ensure efficient water flow.

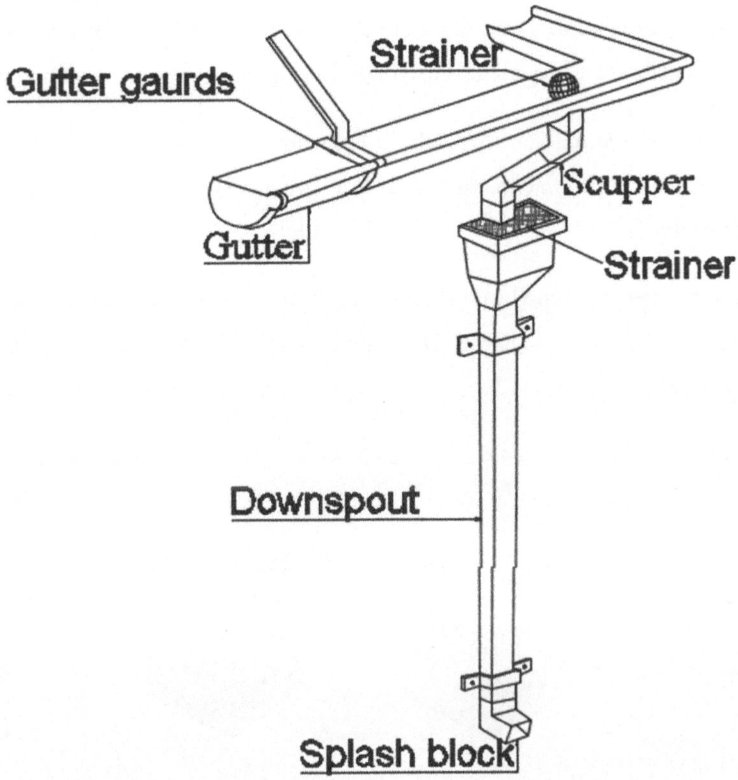

Fig. 4.10: Roof drainage plan

4.8 DEPICTING JOINERY

The Joinery is an important part of woodworking, which is joining wood parts to form objects or structures. The following are some typical joinery types.

(a)Butt Joint: The common type of wood joint it involves more than one piece of wood is placed together. This joint is not a critical component, rather it is used for low strength.

(b) Miter Joint: The ends of the wood are cu at 45^0 angles and joined together. This joint is used in preparing boxes, decorative elements and picture frames.

(c) ***Dovetail Joint:*** The wooden pieces are interlocked with a series of pins and tail, creating a strength joint. This joint is used in furniture's, drawers and boxes

(d) Lap Joint: There is an overlap between the wood pieces, used for basic construction.

(e) Groove and Tongue joint: There is a groove on one piece, and a corresponding tongue on the other that slides into the groove. It is used for flooring, paneling and similar applications.

(f) Mortise and Tenon joint: Mortise is a hole and tenon is a projection. The joint is made as one piece with a tenon fits into a mortise. It is a commonly used joint for making doors, windows and chairs.

(g) Biscuit joint: The oval shaped wooden biscuits are inserted into slots cut into the pieces to be joined, glued and clamped together. This type of joint is used for cabinets, tabletops and other large surfaces.

(h)Pocket joint: one piece is drilled with angled holes, and then fastened to another piece using screws.

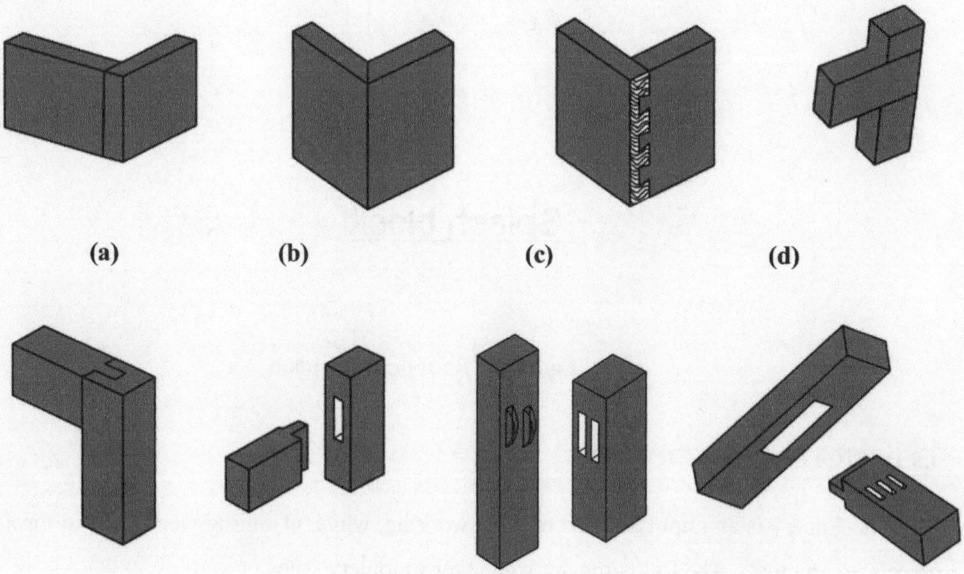

(a) **(b)** **(c)** **(d)**

(e) **(f)** **(g)** **(h)**

Fig. 4. 11: Joints

4.9 STANDARD FITTINGS, FIXTURES AND FINISHES

4.9.1 Fittings

The term standard fittings describe the parts and fittings that are frequently used in a variety of works, such as furniture assembly, plumbing, and electrical work.

(a) Plumbing fittings:

(i) Elbow fitting: A 90^0 or 45^0 pipes fitting that is placed between two lengths of pipe to allow for a direction shift.

(ii) Tee fitting: A T-shaped fitting with three openings that allow a pipe to branch in two directions.

(iii) Coupling: A short length of pipe used to join two pipes.

(iv) Union: An assembly that connects two pipes and permits unplugging without cutting.

(v) Reducer: A fitting is used to join two pipes of various sizes.

Fig. 4. 12: Plumbing fittings

(b) Electrical fittings:

(i) Switch: A device for connecting and disconnecting the electric circuits.

(ii) Socket outlet: It is a device that permits the connection of electrical equipment to the main power source.

(iii) Junction Box: The electrical connections enclosed to keep them safe and avoid short circuits.

Fig. 4.13: Electrical fittings

(c) Furniture fittings:

(i) Corner bracket: A bracket made of plastic or metal that is used to strengthen corner joints while assembling furniture.

(ii) Cam lock: A kind of fastener used to join two pieces of wood in flat-pack furniture.

(iii) Dowel pin: A long, cylindrical rod that strengthens the joint.

(iv) Drawer slide: A device that allows drawers to be opened and closed smoothly.

(v) Shelf support pin: A small pin used to support a shelf within a cabinet.

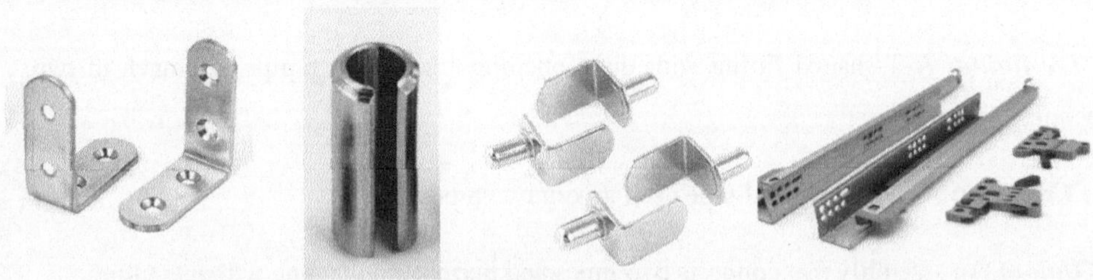

Fig . 4.14: Furniture fittings

4.9.2 Fixtures

In buildings, residences and other structures fixtures are installed permanent or semi-permanent features that are frequently necessary for both functionality and aesthetics. The following are the various types of fixtures

(a) Plumbing Fixtures:

(i) Sink: A sink that's usually found in kitchens and bathrooms and used for hand washing, dishwashing or other functions.

(ii) Toilet: A sanitary fixture for the disposal of human waste.

(iii) Faucet: It is a device for controlling the flow of gas or liquid through a pipe or container.

(iv)*Shower:* A bathing area including a water-spraying nozzle.

(v) Bathtub: A huge container that holds water and can be used for bathing.

Fig. 4.15: Plumbing fixtures

(b) Electrical fixtures:

(i) Ceiling fan: A mechanical fan installed in a room's ceiling that circulates air to make it more comfortable.

(ii) Thermostat: A device that controls a system's temperature to keep it close to a specified set point.

(iii) *Circuit breaker:* It is an electrical switch that operates automatically to guard against overload or short circuit damage to a circuit.

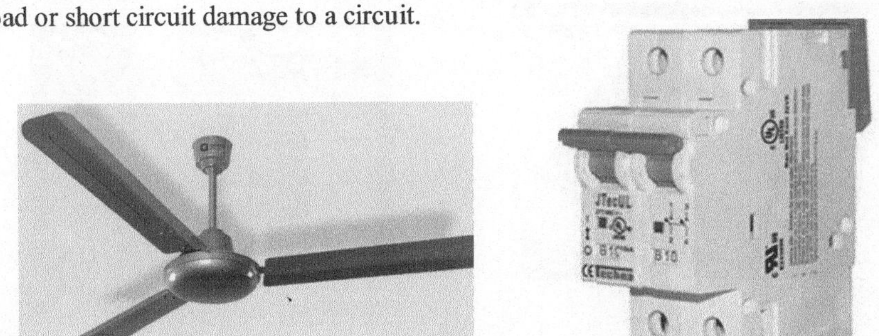

Fig. 4.16: Electrical fixtures

(c) Lighting fixtures:

(i) Table lamp: A portable lamp that rests on a desk or table.

(ii) Pendent light: It is a single ceiling-mounted light fixture, usually supported by a metal rod, chain or cord.

(iii) Chandelier: A hanging light fixture that can hold multiple candles or light bulbs and has branches for decoration.

Fig. 4.17: Lighting fixtures

(d) Kitchen fixtures:

(i) Stove: It is a cooking device, usually run-on gas or electricity.

(ii) Microwave: An appliance used for heating food quickly using microwave radiation.

(iii) Dishwasher: An appliance that cleans dishes automatically.

(iv) Refrigerator: A device that keeps food and beverages cold.

Fig. 4.18: Kitchen fixtures

4.10 USE OF NOTES TO IMPROVE CLARITY

The drawing includes notes that specify the materials, ratios, types, and gauges used in the construction. Drawing notes is very useful and necessary for an engineer's successful understanding of the drawing on site. The following are the general notes provided in the drawing

NOTES - ENGINEERING DEPARTMENT

1. Code of Specification: All material and works shall conform to Indian Standard Specifications (Works & materials)

2. Cement Concrete in Foundation shall be as CC (1:3:6)

3. Substructure: Coursed Rubble masonry in CM (1:6) below GL to PL

4. Plinth Filling: Earth (other than BC soil), filling well-watered and consolidated to required level, that may be followed by Sand Filling and PCC (1:3:6)

5. DPC: 40mm thick. Damp proof course should be provided below full width of wall at Floor Level.

6. Superstructure: ACC Block masonry in CM 1:6 for full brick wall and 1:4 for half brick wall should be used. Main wall thickness shall be 230mm and partition wall of 115mm.

7. RCC Work: Column, Beam, Roof, Chajja, Lintel should be provided with M25 nominal mix as per guidelines and for all other provisions IS: 456-2000 shall be followed

8. Roof Height: Height of ceiling above floor level is 3.75m.

9. Water Proofing of RCC slab & all external walls shall be done by adding water proofing admixtures @ 1Kg per bag of 50Kg of cement.

10. Plastering: All inside wall should plaster with CM 1:6, thickness of plaster should be 12/15 mm. All outside wall should plaster with CM 1:5, thickness of plaster should be 15/20 mm.

11. Flooring: Vitrified tiles of approved brand should be provided in ADEN Chamber, Office, Rest Room &Verandah should be provided with mat finished anti-skid Vitrified tiled. Toilet should be provided with Anti -skid ceramic tiles. Vitrified tiles skirting of 125mm Ht. should be provided in all rooms except toilet. Store Room May be provided with Kota Stone.

12. Dado: Glazed Vitrified /Ceramic tiles of Height 2100mm above floor level should be provided in Toilet.

13. Painting: Outside plaster exposed to weather condition shall be painted with Emulsion Paint & Inside plaster shall be painter with distemper of 2 coats. MS grill should be painted with 2 coats enamel paint. Teak Wood door frame should be polished and painted with 2 coat primers.

14. Rain water pipes of 110mm dia. should be provided for the terrace. Down/rain water pipes and waste water pipes should be of PVC. Necessary arrangement to be made at floor level to avoid rain water entering into rooms.

15. The sill level of window should be 900mm above floor level.

16. PVC doors to be used for bath /WC.

17. Windows should have aluminum frame with sliding type glass shutter, protected by m.s grill & fly proof mesh.

18. The elevation of the building should be designed aesthetically by architecture and got approved by division before commencement of work.

NOTES - ELECTRICAL DEPARTMENT

1. Sufficient illumination shall be ensured inside and outside of Office building with LED lightening.

2. Sufficient number of fans should be provided in as per requirements.

3. Power plug switch board should be provided in all rooms.

4. Electrical wiring should be concealed, as MDS switches to be provided.

5. Ensure shifting of cable and other underground/overhead installation in the proposed site before taking up building work.

6. Necessary earthing arrangements shall be provided such that earth resistance is maintained with in stipulated limits and properly planned with s & t so that the inter distances are maintained.

7. Only MCCBS/MCBS shall be provided in all the main distribution boards and sub distribution boards.

8. GI poles to be provided on circulation/outside area.

NOTES - GENERAL

1. All dimensions are shown in Meters unless otherwise mentioned.
2. Only written dimensions shall be followed.
3. The building proposed as a framed structure and will be designed for G+2 floor.
4. Depth & width of foundation will be decided by design calculation, on basis of nature and SBC of soil. Same will approved by HQ organization as per norms.
5. For Location Plan refer to Drawing number: DRG/No/Year

UNIT SUMMARY

The most useful and needful information for developing architectural to structural drawings is discussed in this unit. The points to remember are

- All the technical terms used in preparing drawings
- Listing out all the various elements required before developing the plan.
- Setbacks are required as per the norms issued by the government.
- Positioning various rooms as per Vaastu and accommodating them in the given area.
- Placing doors, windows, and stairs with appropriate sizes and shapes to provide sufficient ventilation and access to all the rooms
- Positioning of columns without disturbing the aesthetics of the building and selecting a foundation based on the soil strata
- Providing enough surface drainage to drain all the rainwater properly, not affecting the durability of the structure.

EXERCISES

A. Multiple Choice Questions

1. Which of the following is a structural element of a building?

   a) Door

   b) Window

   c) Footing

   d) All the above

2. ______is known as a compression member.

   a) Beam

   b) Column

   c) Wall

   d) Slab

3. General thickness of a partition wall

   a) 2”

   b) 8” & 9”

   c) 4 ½” or 6”

   d) 3” or 4”

4. ______is a joint that overlaps between the wood pieces.

   a) Biscuit joint

   b) Dovetail joint

   c) Miter joint

   d) Lap joint

5. What is the purpose of gutter channels in buildings?

   a) To collect rainfall

   b) To use as a sewer

   c) To use as a waterline

   d) None of the above

Answers

1. c   2. b   3. c   4. d   5. a

B. Short and Long Answer Type Questions:

4.1 Explain briefly about the following terms

a) Carpet area

b) Plot ratio

c) Gross Floor area

d) Plinth area

4.2 What are elements of Planning and building drawing?

4.3 Explain about Structural Elements.

4.4 What is meant by roof drainage plan? And Explain about its key components.

4.5 Explain in step-step of preparing detailed plan.

KNOW MORE

- *Parking Studies:* There is more scope for students not to get restricted with simple houses and to get in-depth knowledge for planning higher-storey buildings, which require parking as mandatory.
- *Rainwater Harvesting:* Proper tools and techniques can be studied and developed for effectively using the drained water from rains in the monsoon season. This water can be treated and utilized for households throughout the year if the system is planned effectively.
- *Sustainable Materials:* Engineers/architects should have a thorough knowledge of all the materials to be used as sustainable materials in the construction. The use of sustainable materials in the construction will bring a positive impact on the cost and durability of the building, making research on cost-effective durable materials is the need of the present scenario.
- *Modern Tools:* Earthen tunnels, natural ventilation, etc. developed for greener construction, which reduces the use of fuels and creates a positive impact on the environment. Developing such tools for effective living creates a friendly environment and utilizes non-renewable resources.

REFERENCES AND SUGGESTED READINGS

1. Dr. N. Kumara Swamy and A. Kameswara Rao. Building Planning and Drawing, 9th ed.; Charotar Publications, 2023.
2. S.S. Bhavikatti and M.V. Chitawadagi. Building Planning and Drawing, Wiley Publications, 2019.
3. B.P. Verma. Civil Engineering Drawing & House Planning, Khanna Publishers, 2023.
4. National Building Code of India 2016.

Dynamic QR Code for Further Reading

5 PICTORIAL VIEW

UNIT SPECIFICS

The following aspects are discussed in this module.

- Isometric views
- Terms in perspective views
- Types of perspective views
- How to draw a perspective view of a building
- Building Information Modeling (BIM)

RATIONALE

This chapter makes the students learn the basics of 3D drawings available. The student will learn how to draw the 3D projections of a building effectively.

PRE-REQUISITES

- Elements of a building
- Basic Knowledge of projections
- Stages of Construction

UNIT OUTCOMES

The following are the outcomes of this unit:

U5-O1: Understand the basic terms of projections.

U5-O2: Identify the views of a building from different angle of visions and positions.

U5-O3: Understand the fundamentals of building information modelling

<table>
<tr><th rowspan="2">Unit- 5
Outcomes</th><th colspan="5">EXPECTED MAPPING WITH COURSE OUTCOMES
(1-Weak correlation; 2-Medium correlation; 3-Strong correlation)</th></tr>
<tr><th>CO-1</th><th>CO-2</th><th>CO-3</th><th>CO-4</th><th>CO-5</th></tr>
<tr><td>U5-O1</td><td>1</td><td>3</td><td>3</td><td>2</td><td>1</td></tr>
<tr><td>U5-O2</td><td>2</td><td>3</td><td>3</td><td>3</td><td>2</td></tr>
<tr><td>U5-O3</td><td>3</td><td>2</td><td>1</td><td>1</td><td>3</td></tr>
</table>

5.1 ISOMETRIC VIEW

It is a method for visually representing three dimensional objects in two dimensional in technical andengineering drawings. It is an axon metric projection which the three coordinate axes appear equally for short end. The angle between any two axes is 120 degrees

5.1.1: Isometric Axes

The three lines OA, OB and OC meeting at a point O and making the angles of three 120^0 with eachother are known as isometric axes

5.1.2: Isometric Drawing

The 3D View of an object drawn with axes as shown below with true dimensions is called isometric view or isometric drawing. All the vertical lines are always perpendicular to the X-axis and all other horizontal lines are always inclined at 30^0 to the horizontal.

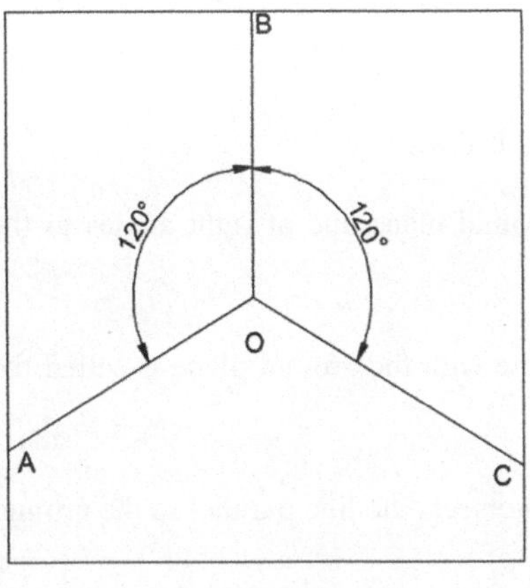

Fig. 5.1: Isometric axes

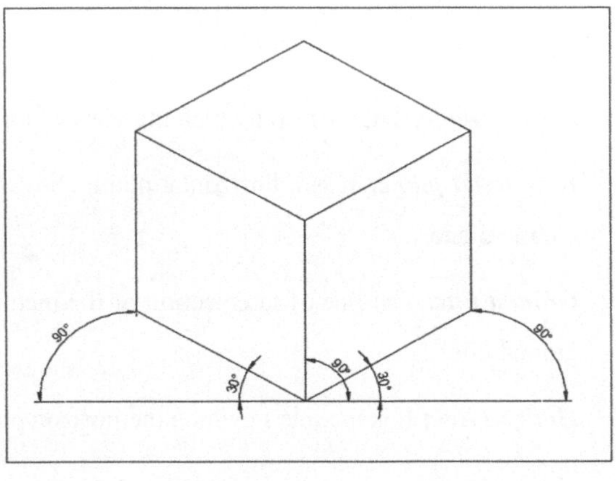

Fig. 5.2: Isometric View

5.2 PERSPECTIVE VIEW

The realistic view of an object perceived by the eye of an observer is known as perspective of a projection. The perspective view of a building shows the actual projection of the building in different orientations and eye positions. The principle of perspective view depends on conical projections.

5.2.1: Principle of Perspective Projections

Perspective is a fundamental concept in visual arts that helps the artist to create the illusion of depth andspace on a two dimensional surface. It is a graphic resource that uses converging lines and other techniques to make objects appearcloser or farer away creating a sense of depth and dimensionality.

5.2.2: Terms in Perspective Projections

Ground plane: It is a horizontal plane on which the object is assumed to be situated.

Station plane: It is the point where the eye of the observer is located while viewing the object.

Picture plane: The plane on which the view of an object is drawn.

Horizontal plane: It is a horizontal plane abovethe ground plane and at right angles to the picture plane.

Ground line: The line of intersection of the picture plane with the ground plane is called the ground line

Horizon line: It is the line in which the horizon plane intersects the line parallel to the ground line.

Perpendicular axis: It is the line drawn through the station point perpendicular to the picture plane. It is also called as the line of sight or axes of vision

Station point: The point from where the observer is looking at the object or building. The size and shape of the object depends on the position of station point.

Angle of vision: Angle of vision is the rotation angle of the object with respect to picture plane.

Eye level / Eye point: It represents the height between the ground line and the eye point of the observer.

5.2.3: Methods of Drawing Perspective view

(a) Visual Ray Method

Visual ray method is a fundamental concept in perspective projections, particularly in thefield of computer graphics and architectural design. This method involves tracing rays from the viewer's eye (or the camera) through each pixel on the viewing plane and determining where these rays intersect with objects in the scene. This intersection point gives the location of the corresponding pixel on the object's surface. By repeating this process for each pixel, a complete image of the scene from the viewer's perspective can be generated.

(b) Vanishing Point Method

In perspective projections, a vanishing point is a point on the horizon line where parallel linesin the scene appear to converge or "vanish" when extended into the distance. These points are crucial for creating the illusion of depth and distance in a two-dimensional representation of a three-dimensional scene.

Vanishing points are a key aspect of linear perspective, a technique used in art and architecture to create the appearance of depth and distance on a flat surface. Artists and designers use vanishing points to accurately depict how objects and lines recede into the distance, creating a sense of perspective and realism in their work.

In simple terms, the vanishing point is where parallel lines appear to meet in the distance when observed from a specific viewpoint. The number of vanishing points in a scene depends on the number of sets of parallel lines that exist within the scene. In one-point perspective, there is typically only one vanishing point, while in two-point perspective, thereare two vanishing points, and in three-point perspective, there are three vanishing points.

(i) One point perspective (or) parallel perspective

One-Point Perspective Also known as frontal perspective, this type of perspective uses a single vanishing point to create the illusion of depth and distance. It's often used for images with a simple frontal view, such as architectural drawings of buildings.

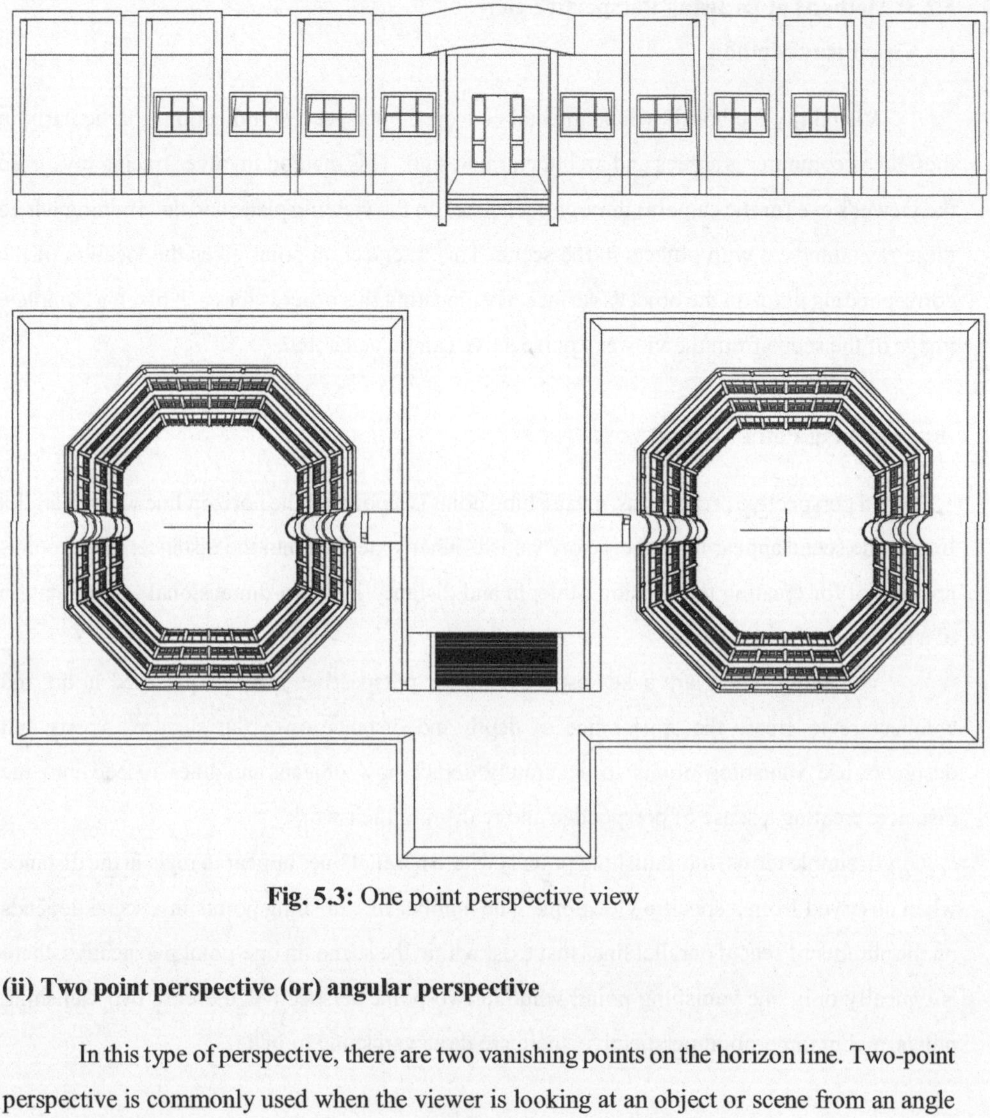

Fig. 5.3: One point perspective view

(ii) Two point perspective (or) angular perspective

In this type of perspective, there are two vanishing points on the horizon line. Two-point perspective is commonly used when the viewer is looking at an object or scene from an angle rather than straight on. It's often used for drawings of buildings or interiors.

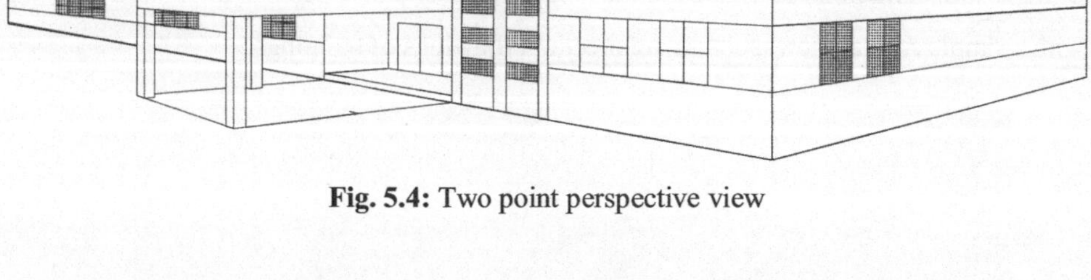

Fig. 5.4: Two point perspective view

(iii) Three point perspective (or) oblique perspective

Three-point perspective adds a third vanishing point, usually above or below the horizon line, to account for the vertical lines of objects that are tilted away from the viewer. This typeof perspective is useful for drawing objects viewed from extreme angles, such as looking up at tall buildings or down at deep canyons.

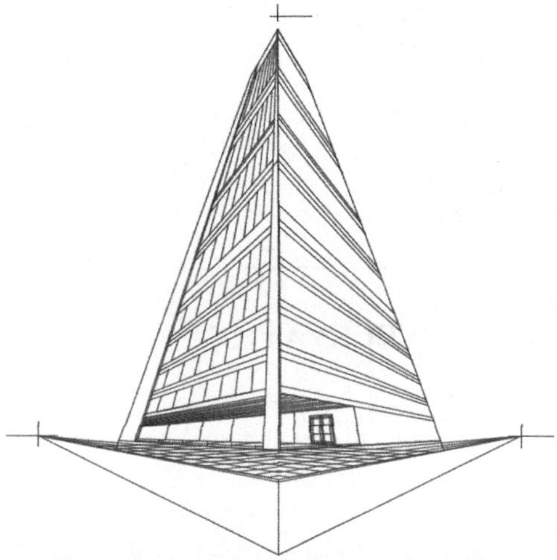

Fig. 5.5: Three point perspective view

5.2.4: Drawing Perspective view of Building

Draw the perspective view of a single room building shown in figure 5.6. Assume station point and ground level is 10.0m from the picture plane. Eye point is 1.5m

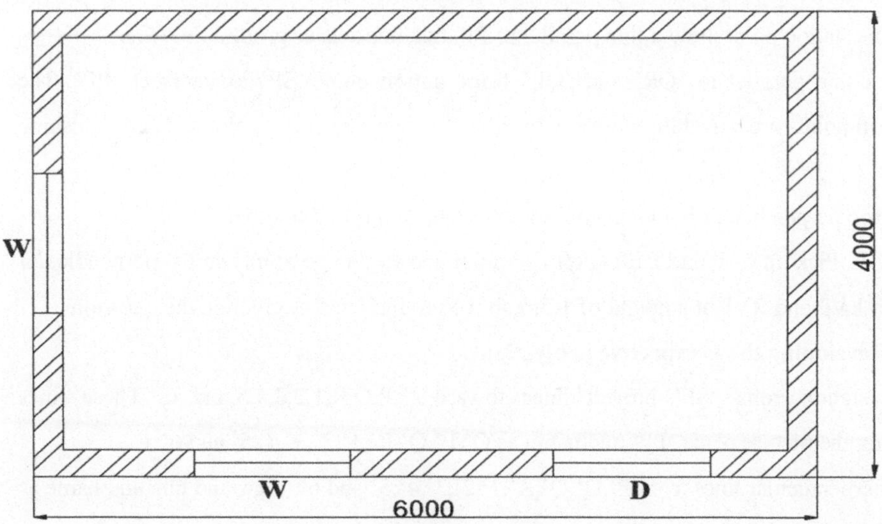

PLAN

above ground level. Assume 30^0 as angle of vision.

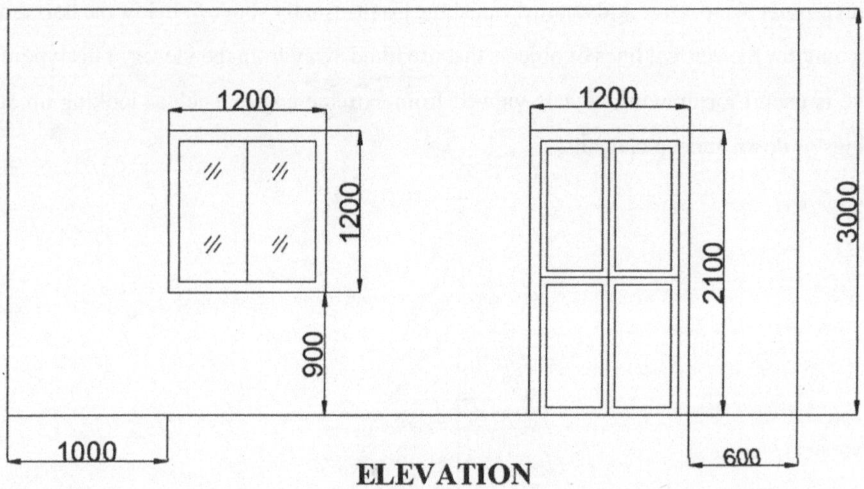

Fig. 5.6: Plan and elevation of a single room

Part A: Drawing all the planes and points

1. Draw an imaginary horizontal line and name it as 'PP' (Picture Plane)
2. Draw the plan given with 30^0 inclination to the PP.
3. Drop the reference point "O" downward perpendicular to PP for a distance of 10.0m, the point obtained is station point 'SP').
4. From the station point draw a line parallel to PP, this line acts as ground line 'GL'.
5. Draw a line parallel to 'OR' and 'OP' from station point 'SP' to intersect 'PP'. The intersection points are vanishing points, (RVP, LVP) right vanishing point and left vanishing point.
6. Now drop perpendicular points from LVP and RVP on ground line 'GL'.
7. From GL, Pick up LEP and REP (Left eye point and Right eye point) on the perpendicular lines from LVP and RVP at a height of 1.5m above ground level as given in the question.

Part B: Developing the perspective projection

1. From station point 'SP' project lines toward O,P,Q,R,1,2,3,4,5,and 6. These lines intersecting the picture plane 'PP' and name as O',P',Q',R',1',2',3',4',5' and 6'.
2. Draw perpendicular lines from P',O',Q',R',1',2',3',4',5' and 6' to ground line and name as P".O",Q",R",1",2",3",4",5"and 6".

3. From station point 'SP' project lines from SP to REP and LEP and mark the points as P_0,5_0,6_0 and R_0 on the respective perpendicular lines P'P", 5'5",6'6" ad R'R''.

4. Draw a line SP-O_1 with a length of 3.0m (from elevation) from station point.

5. Draw a line projecting from O_1 to REP and LEP and mark the points P_1, O_1, R_1 on P'P", O'O", R'R".

6. Join the lines P_0P_1, O"O_1 (or) SP-O_1, R_0R_1 with thick lines. These lines indicate the outer lines of the building.

7. Mark the points L_1 and L_2 on SP-O_1 at a height of 2.1m and 0.9m representing the top point of doors and bottom point of windows.

8. Draw lines from L_1 and L_2 to REP and LEP. These lines intersect the perpendicular 5'5", 6'6", 3'3",4'4",1'1",2'2" at $5_1 5_0$, $6_1 6_0$,$3_1 3_0$,$4_1 4_0$,$1_1 1_0$,$2_1 2_0$ connect all these lines with thick lines, it indicates the projections of doors and windows.

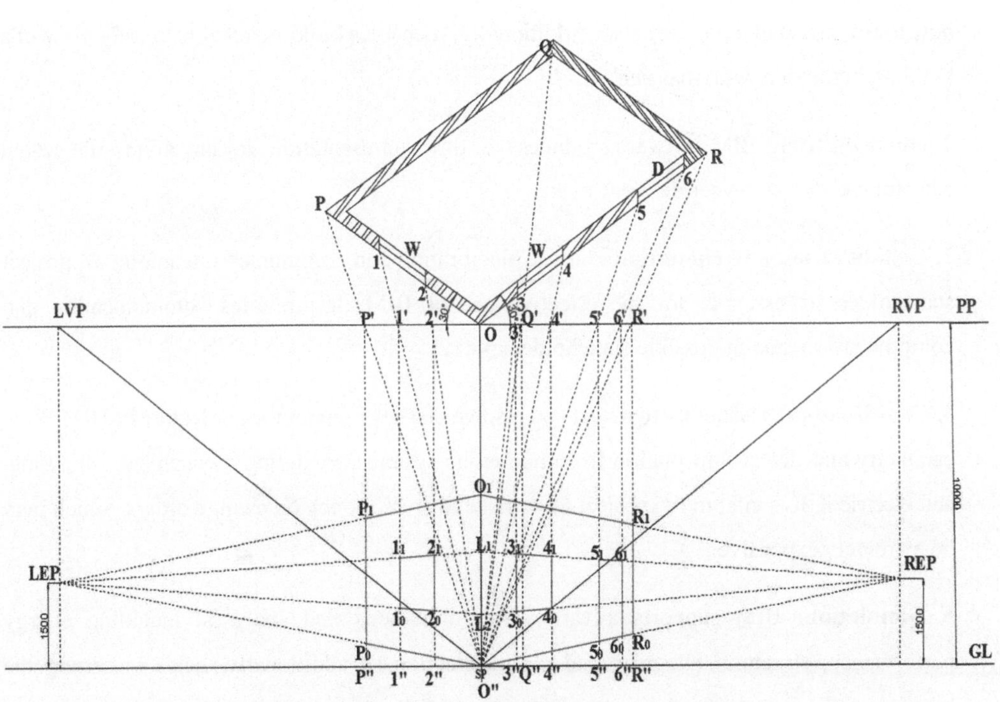

Fig. 5.7: Perspective view of single room building

5.3: FUNDAMENTALS OF BUILDING INFORMATION MODELING (BIM)

Building Information Modeling (BIM) is an intelligent 3D model-based approach that provides architecture, engineering, and construction professionals with knowledge and resources to enable more efficient planning, design, construction, and management of buildings and infrastructure. BIM aims to improve project results by enhancing coordination, eliminating errors, and allowing for more informed decisions throughout the project lifecycle.

Components of BIM

1. **Geometry:** The 3D model of the building

2. **Data:** Specifications regarding materials, costs, schedules, and performance.

3. **Visualization:** Realistic rendering and walkthroughs

4. **Cooperation:** Access shared by all parties involved.

Fundamentals of BIM

1. **Visualization:** BIM generates a 3D virtual model of a project, allowing owners to visualize how a structure will be constructed. Additionally, it enables building owners to fully view of a building before any work has started.

2. **Quantification:** BIM software produces real-time representations for any minuscule design adjustments that an owner desires t make.

3. **Collaboration and communication:** Collaboration and communication among all project stakeholders is essential for the effective use of BIM. It promotes communication and comprehension among possible stakeholders.

4. **Coordination:** The ability to detect and resolve clashes is an important feature of BIM. BIM can study and detect information from numerous systems, including mechanical, plumbing, and electrical. It is vital to the project team since it saves money on change orders, which may be extremely expensive.

5. **Simulation:** BIM supports a variety of simulations and analyses, including energy performance, structural integrity, and clash detection. It additionally, helps in setting the schedule and preparing for the construction phase of the project.

UNIT SUMMARY

Drawing/developing plans is not only the biggest challenge, but developing various views in 2D and 3D is the most important challenge for architects/engineers. This chapter deals with developing various 3D views. The points to be remembered are.

- Adjusting various view positions depending on the requirements of the client.
- Learning all the technical teams of 3D views and orientations.
- Developing isometrics and perspectives by understanding the position of the eye point and inclination with the ground.
- Tools used in building information modelling
- *Developing/generating models in BIM for effective monitoring of construction &* execution.

EXERCISES

A. Multiple Choice Questions

1. The angle difference between the true axes and isometric axes

a) 30^0 b) 45^0 c) 15^0 d) 90^0

2. Which of the projection is mainly used in engineering perspective?

a) Isometric projection b) Perspective projection c) One point projection d) None of the above

3. In isometric projection all the three principal axes are inclined at an angle of

a) 30^0 b) 45^0 c) 60^0 d) 120^0

4. The isometric projection of a circle is

a) Ellipse b) Circle c) Sphere d) Cube

5. The isometric view is the drawing with

a) Isometric scale b) Vernier scale c) Actual scale d) Reduced scale

Answers

1. a 2. a 3. d 4. c 5. c

B. Short and Long Answer Type Questions

5.1 What is Isometric View?

5.2 What is Perspective view?

5.3 Define RVP and LVP?

5.4 What is Angle of vision?

5.5 What are the types of perspective views?

5.6 Define BIM?

5.7 What are the fundamentals of BIM?

KNOW MORE

- *Revit:* It is a software tool for developing the real-time building information model. Revit is exceptionally used now a days for effective control over the execution.
- *3DArchitechre /sketch up:* are the tools for developing isometric & perspective views apart from CAD. These tools are very effective in developing viewpoints based on the position & orientation of the observer.

REFERENCES AND SUGGESTED READINGS

1. N.D. Bhatt. Engineering Drawing, 53rd ed.; Charotar Publishers, 2017.
2. Robert W. Gill. Perspective from Basic to Creative, Thames & Hudson Publications, 2006.
3. Dr. Adv. Harshul Savla, Dr. Chandrahauns Chavan and Ar. Palavi Patil. Building Information Modeling, Notion Press, 2021.

Dynamic QR Code for Further Reading

CO AND PO ATTAINMENT TABLE

Course outcomes (COs) for this course can be mapped with the programme outcomes (POs) after the completion of the course and a correlation can be made for the attainment of POs to analyze the gap. After proper analysis of the gap in the attainment of POs necessary measures can be taken to overcome the gaps.

Table for CO and PO attainment

Course Outcomes	**Expected Mapping with Programme Outcomes** *(1- Weak Correlation; 2- Medium correlation; 3- Strong Correlation)*											
	PO-1	**PO-2**	**PO-3**	**PO-4**	**PO-5**	**PO-6**	**PO-7**	**PO-8**	**PO-9**	**PO-10**	**PO-11**	**PO-12**
CO-1												
CO-2												
CO-3												
CO-4												
CO-5												

The data filled in the above table can be used for gap analysis.

INDEX

Practical Sheets

Practice Sheet No. 1

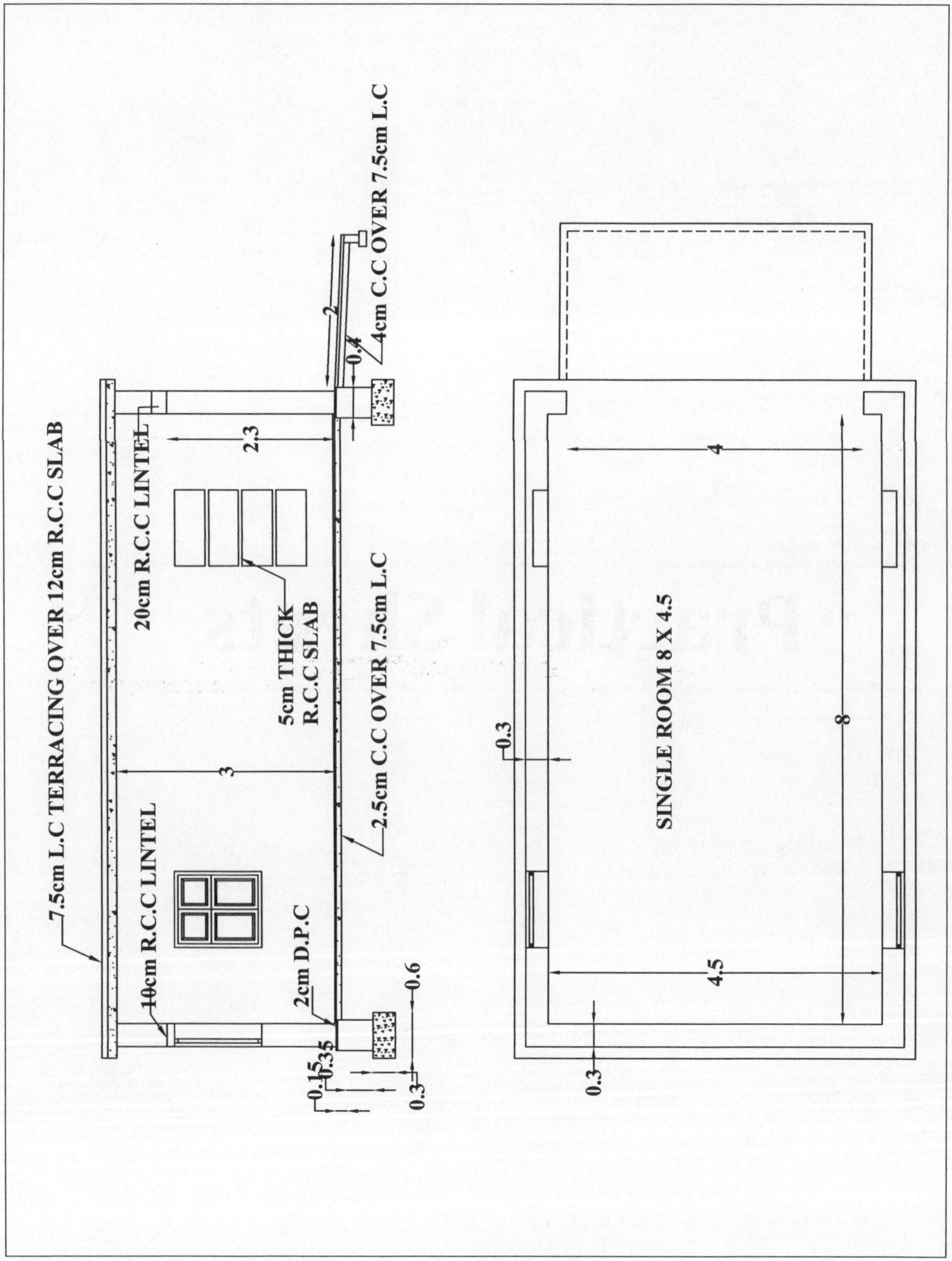

Practice Sheet No. 2

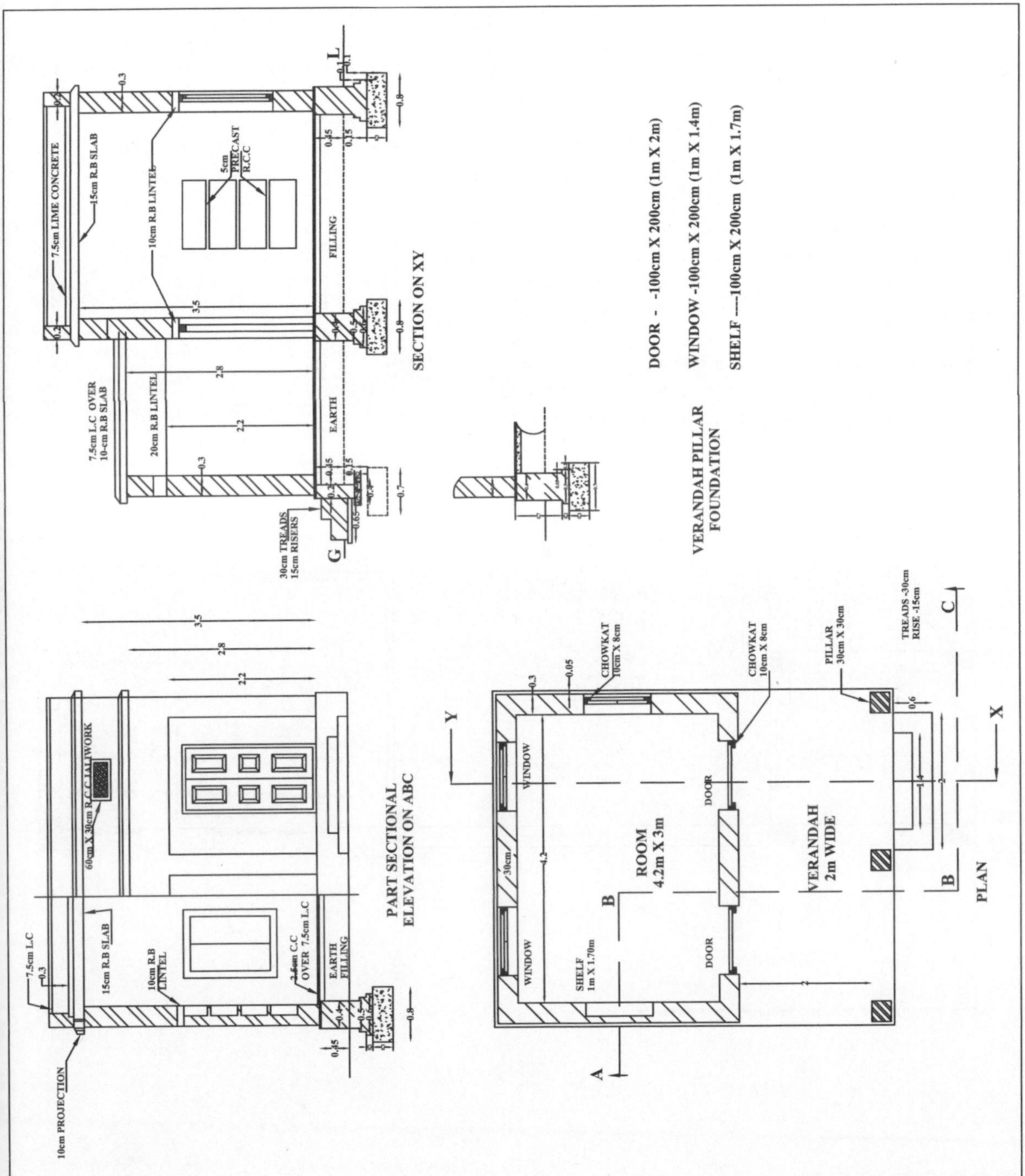

Practice Sheet No.

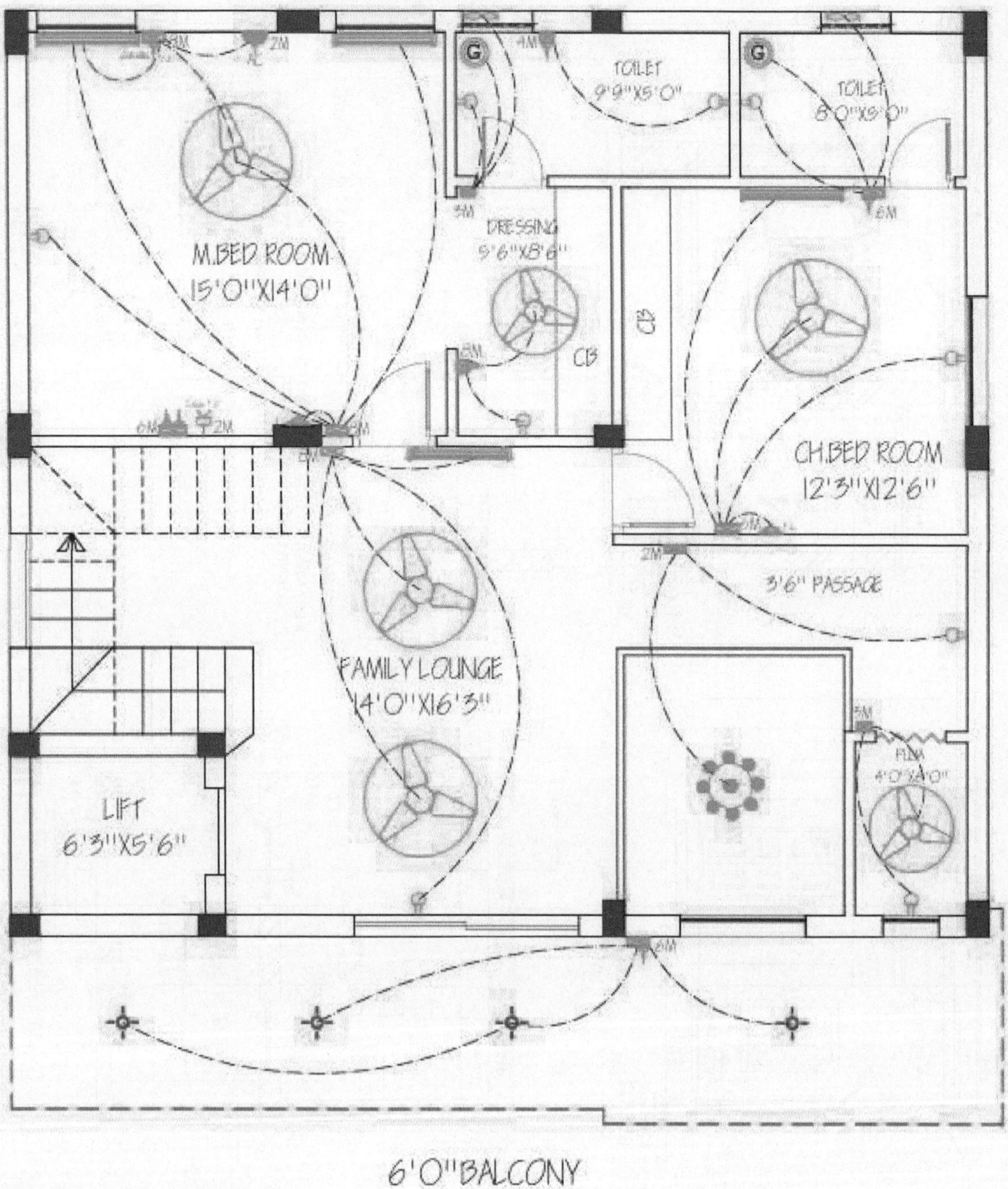

FIRST FLOOR

ELECTRICAL DRAWING

Practice Sheet No. 4

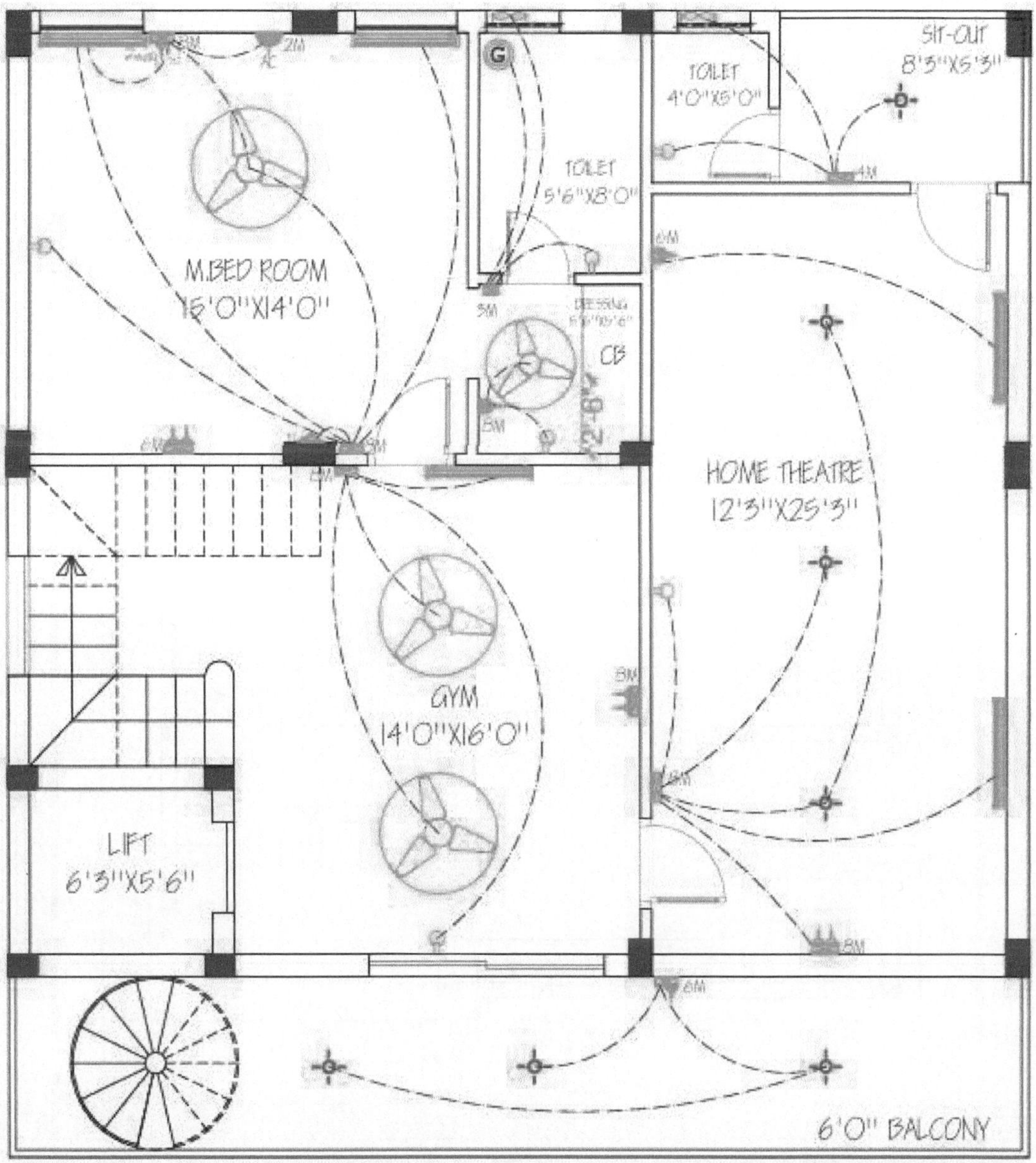

SECOND FLOOR

Practice Sheet No. 5

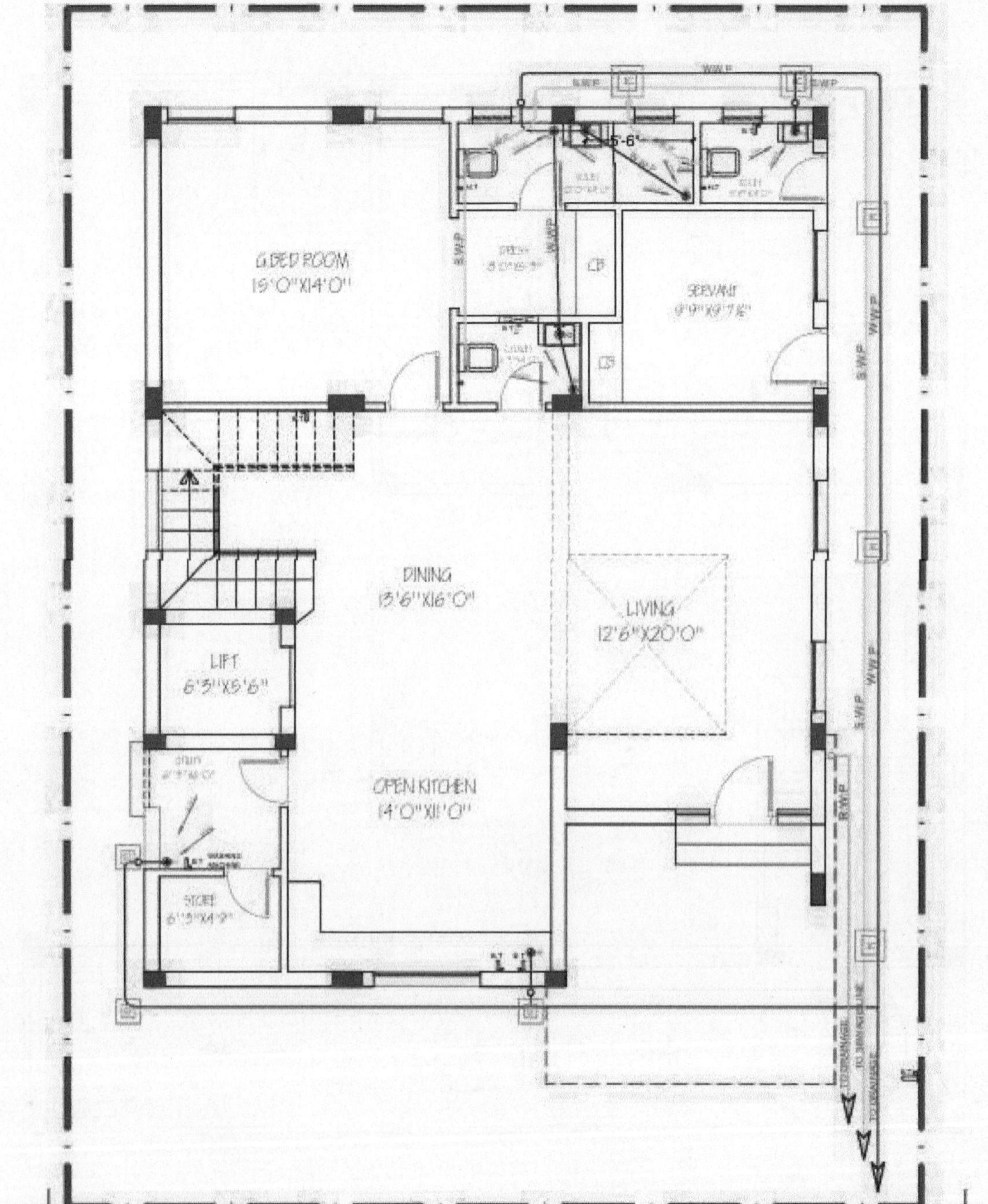

GROUND FLOOR

PLUMBING DRAWING

Practice Sheet No. 6

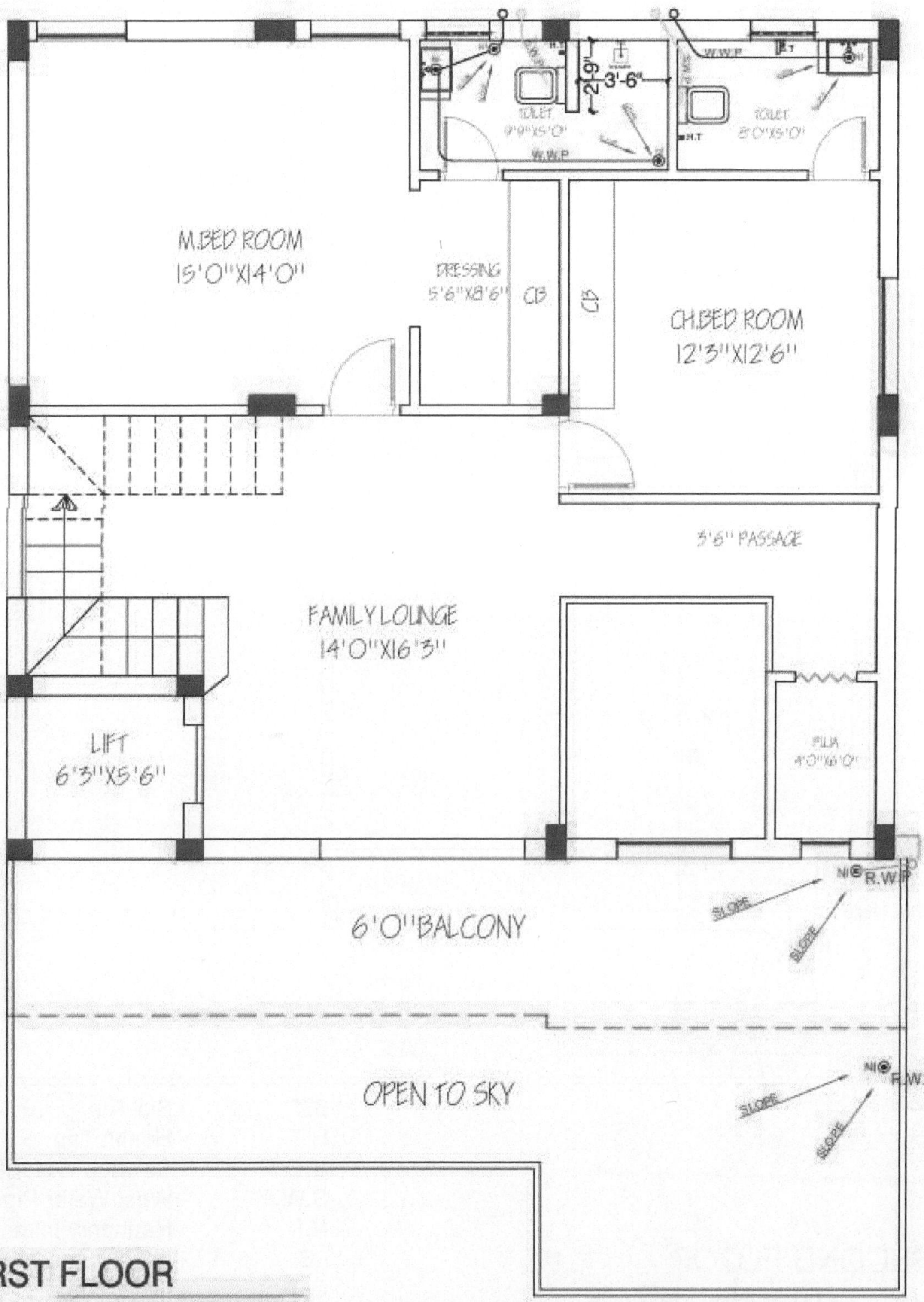

FIRST FLOOR

PLUMBING DRAWING

Practice Sheet No.

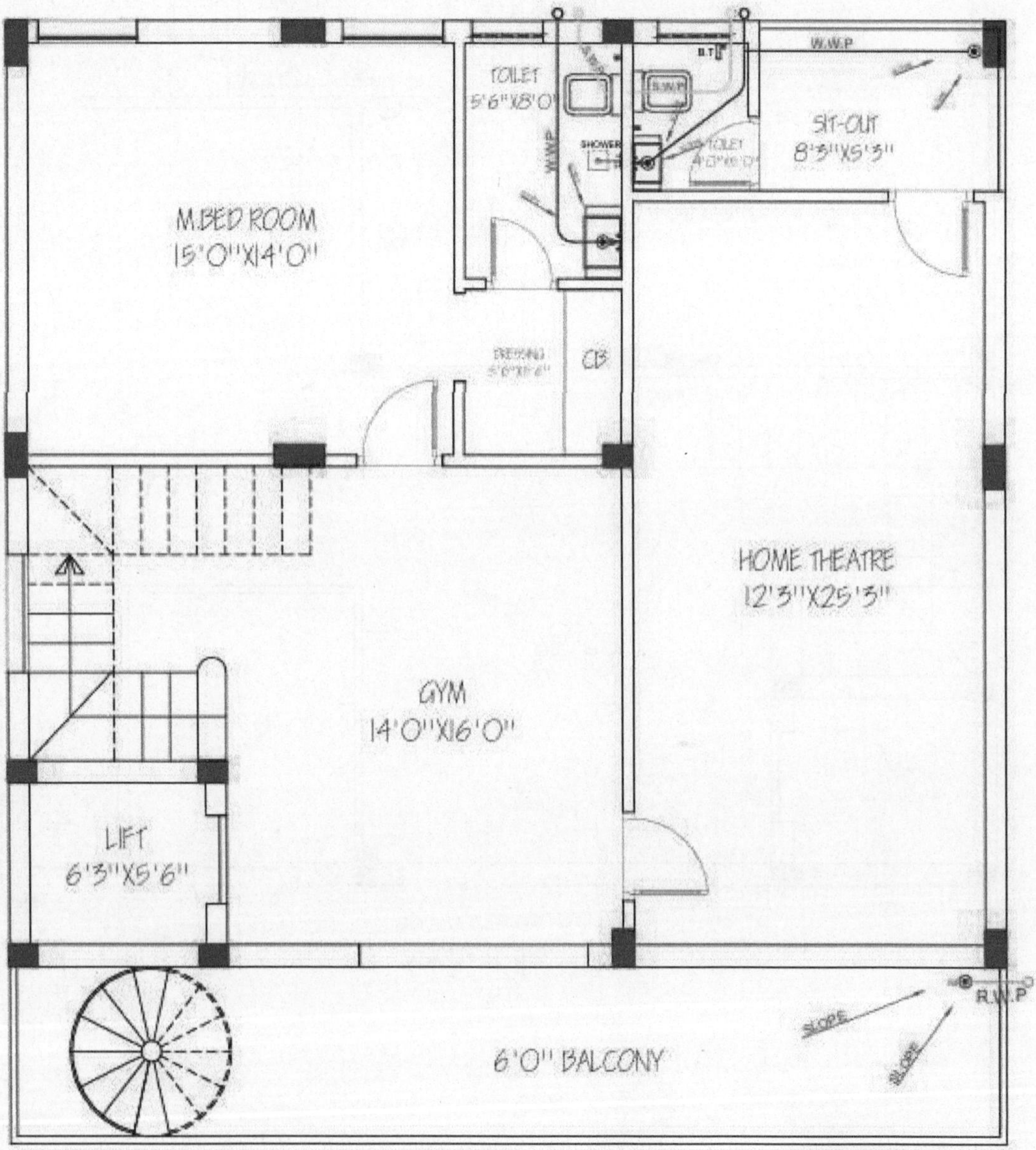

LEGEND

B.T	- Bib Tap
H.T	- Health Tap
S.W.P	- Sewage Water Pipe
S.W.P	- West Water Pipe
N.I	- Nauhenry Inlet
I.C	- Inspection Chamber
G.T	- Gully Trap

SECOND FLOOR

PLUMBING DRAWING

Practice Sheet No. 8

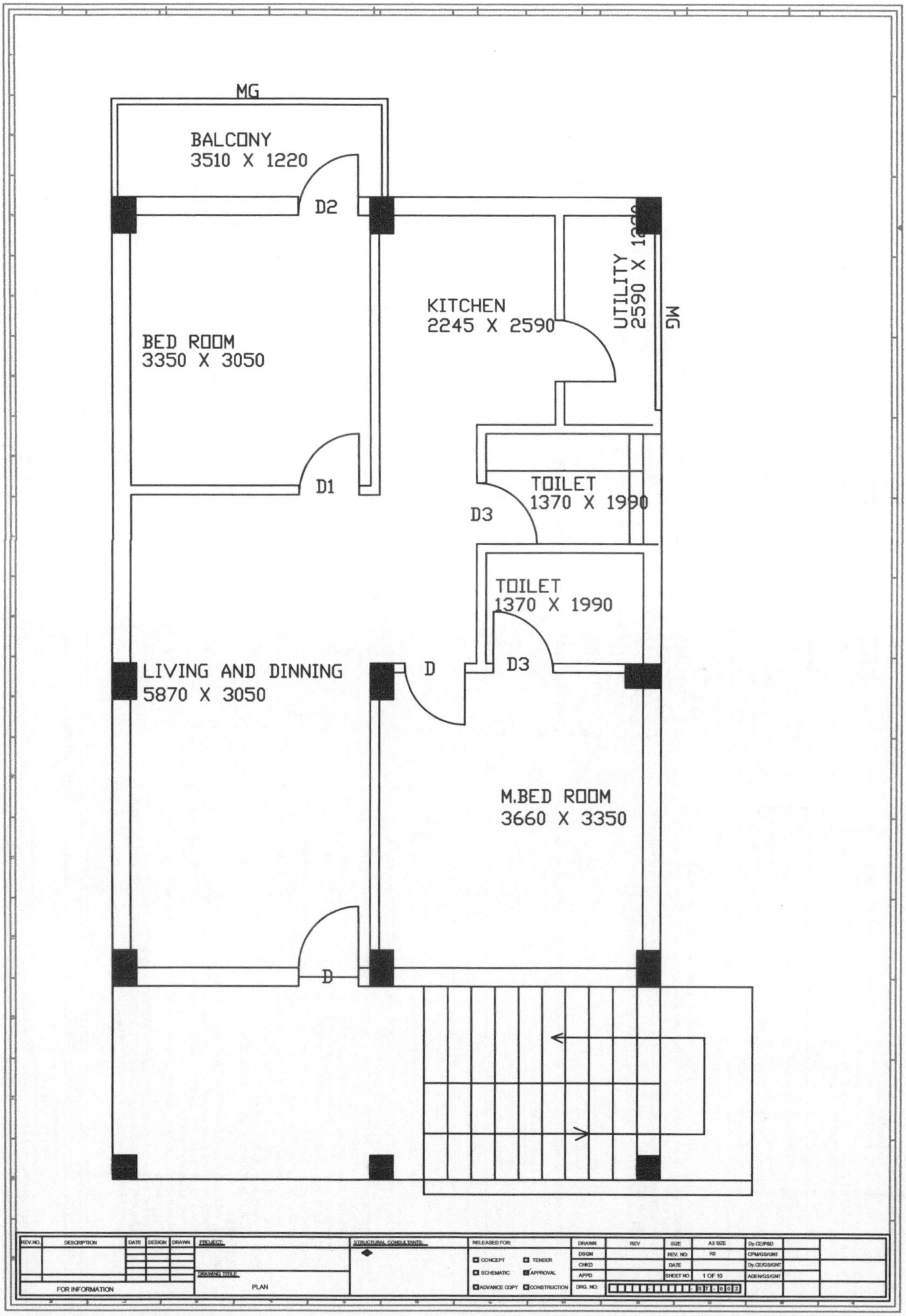

Practice Sheet No.

GENERAL NOTES:

1. DO NOT SCALE THIS DRAWING. FOLLOW FIGURED DIMENSIONS ONLY.
2. ALL DIMENSIONS ARE IN MM ONLY AND LEVELS ARE IN METERS UNLESS SPECIFIED.
3. S.B.C OF SOIL IS CONSIDERED AS 150KN/Sq.mt AS PER SOIL REPORT.
4. USE VIBRATED REINFORCED CONCRETE CONFORMING TO IS:456-2000. THE GRADES FOR DIFFERENT COMPONENTS ARE AS FOLLOWS

COMPONENT	FOOTING	PEDESTAL & COLUMNS	BEAMS	SLABS	WATER TANK SIDE WALLS
MIX	M25	M25	M25	M25	M25

5. GRADE OF STEEL SHALL BE Fe 500 CONFORMING TO IS:1786-2008 & AMENDMENTS.
6. CONCRETE LAID IN THE FORM WORK SHALL INVARIABLY VIBRATE TO OBTAIN MAXIMUM DENSITY AND COMPACTION.
7. PROVISIONS MADE IN IS:456-2000 AND SP:34 AND OTHER RELEVANT CODES SHOULD BE STRICTLY ADHERED TO, DURING EXECUTION.
8. LAPS SHALL BE STAGGERED AND LAP LENGTHS ARE AS FOLLOWS.

MIX	M25
Ld	40Dia

9. NOT MORE THAN 50% OF COLUMN REINFORCEMENT SHALL BE LAPPED AT A PARTICULAR LOCATION.
10. CLEAR COVERS ARE AS FOLLOWS

FOOTING TO MAIN BAR	PEDESTAL TO MAIN BAR	COLUMN TO MAIN BAR	BEAMS TO STIRRUPS	SLABS TO MAIN BAR
60mm	40mm	40mm	25mm	20mm

11. ANY LOOSE OR PROTRUDING BOULDERS SHALL BE REMOVED BEFORE LAYING FOUNDATION CONCRETE AND SHALL BE FILLED WITH PCC (1:4:8).
12. WHEREVER NECESSARY DRILLING SHALL BE RESORTED TO, IN HARD SHEET ROCK ONLY.
13. AT ALL EXTERNAL COLUMN PITS, STONE MASONRY SHALL BE CONSTRUCTED FROM TOP OF FOOTING.
14. THE DIFFERENCE IN LEVEL OF TWO ADJACENT FOOTING SHOULD NOT BE GREATER THAN HALF THE CLEAR DISTANCE BETWEEN THEM.
15. AT THE JUNCTIONS OF INTERNAL PLINTH BEAMS MASONRY PEDESTALS (MP) AT LOCATION SHOWN IN THE PLINTH BEAM DWG. IS TO BE PROVIDED.
16. ALL PLINTH BEAMS SHALL BE CAST ONLY AFTER COMPLETION OF BASEMENT FILLING AND WELL COMPACTION AND LAYING OF PCC BELOW INTERNAL PLINTH BEAMS.
17. STIRRUPS SHOULD HAVE STANDARD HOOK AS PER SP-34
18. EXTRA BARS AT TOP FOR BEAM SHALL BE EXTENDED UPTO 0.25L ON EITHER SIDE FROM THE FACE OF SUPPORTS AND CURTAILMENT OF BOTTOM BARS AT 0.1L FROM DISCONTINUOUS EDGE AND 0.15L FROM CONTINUOUS EDGE MEASURED FROM CENTRE OF SUPPORT, WHERE 'L' IS C/C OF SUPPORTS..
19. SLAB BARS ARE TO BE CRANCKED AT 0.25 L FROM CENTRE OF SUPPORT NEAR CONTINUOUS EDGES AND 0.15 L FROM CENTRE OF SUPPORT NEAR DISCONTINUOUS EDGES.
20. IN SLABS EXTRA BARS OVER CONTINOUS SUPPORT SHOULD BE TAKEN UPTO 0.3L FROM THE FACE OF BEAM. CRANK BARS AT CONTINOUS SUPPORT SHOULD NOT BE TERMINATED ON BEAM AND SHOULD BE EXTENDED UPTO 0.3L FROM FACE OF SUPPORT INTO THE ADJACENT SPAN.
21. DISTRIBUTION STEEL IN SLAB FOR CRANKED BARS AND TOP EXTRA BARS SHALL BE 8 TOR @ 200 C/C
22. THE REINFORCEMENT SHALL BE CHECKED BEFORE LAYING CONCRETE.
23. WHEREVER, THE SECONDARY BEAM RESTS ON MAIN BEAM THE ARRANGEMENT OF BENT UP HANGER BARS SHALL BE PROVIDED AS GIVEN BELOW:

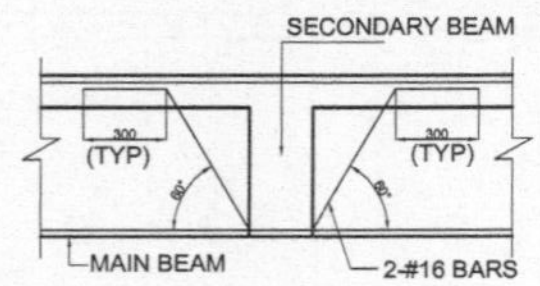

24. IN BEAMS & SLABS, OVERLAPPING SHALL NOT BE PROVIDED OVER SUPPORTS FOR TOP BARS AND AT MID SPAN FOR BOTTOM BARS.
25. WHEREVER REINFORCEMENT IS PROVIDED IN TWO LAYERS IN BEAMS SPACER BAR SHALL BE PROVIDED TO MAINTAIN 25MM GAP OR MAXIMUM DIA OF MAIN BAR,WHICH EVER IS MORE, BETWEEN TWO LAYERS.
26. WHEN SPAN IS MORE THAN 6.0M PROVIDE A CAMBER OF SPAN L/250 FOR BEAMS & SLABS.
27. THIS DRAWING SHOULD BE READ IN CONJUNCTION WITH RELEVANT ARCHITECTURAL DRAWINGS.
28. ALL BUILDINGS ARE DESIGNED FOR WIND SPEED OF 50m/s AND FOR SEISMIC ZONE -3.
29. FOR NUMBER OF FLOORS CONSIDERED IN DESIGN REFER COLUMN C/L DWG OF EACH BLOCK.
30. FOR EXACT LOCATION OF WALLS REFER RELEVANT ARCHITECTURAL DRAWINGS FOR NECESSARY DETAILS.
31. CURING SHOULD BE DONE ATLEAST FOR 21 DAYS.
32. STRIPPING TIME OF FORMWORK SHALL CONFORM TO CLAUSE 11.3 OF IS:456-2000.

33 THIS BUILDING DESIGNED FOR G + 2 FUTURE FLOOR ONLY

REV.NO.	DESCRIPTION	DATE	DESIGN	DRAWN	PROJECT:	STRUCTURAL CONSULTANTS:	RELEASED FOR:	DRAWN	REV	SIZE	A3 SIZE
							☐ CONCEPT ☐ TENDER	DSGN		REV. NO:	R0
					DRAWING TITILE		☐ SCHEMATIC ☑ APPROVAL	CHKD		DATE	
	FOR INFORMATION				GENERAL NOTES		☐ ADVANCE COPY ☐ CONSTRUCTION	APPD		SHEET NO	2 OF 10
								DRG. NO:	- ST - 001		

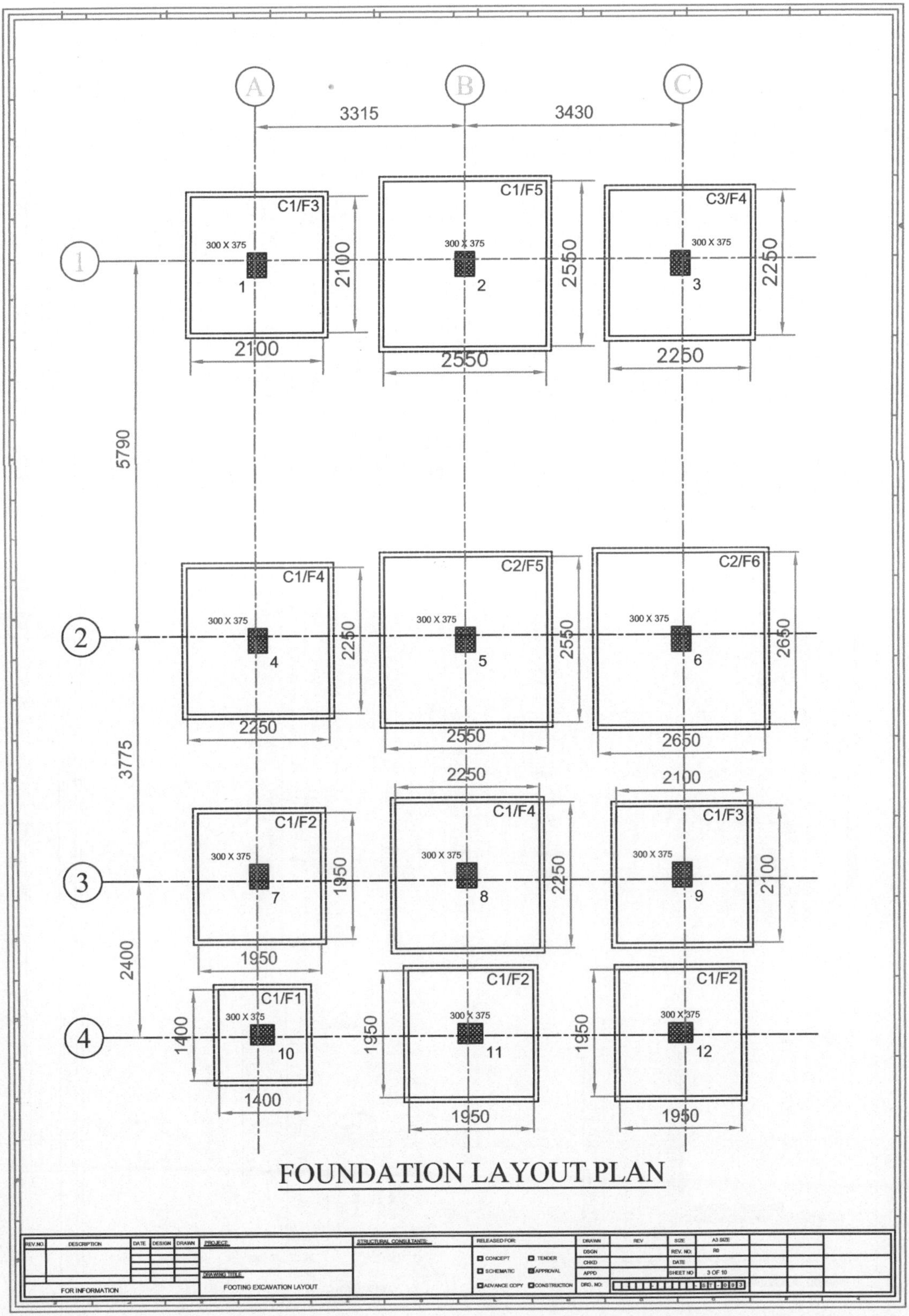

FOUNDATION LAYOUT PLAN

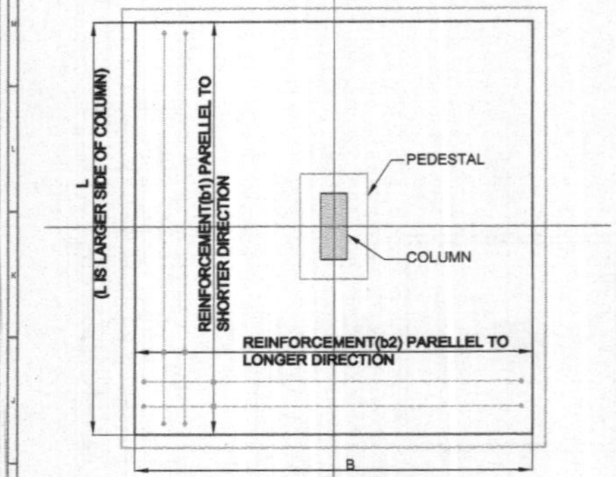

TYPICAL FOOTING PLAN

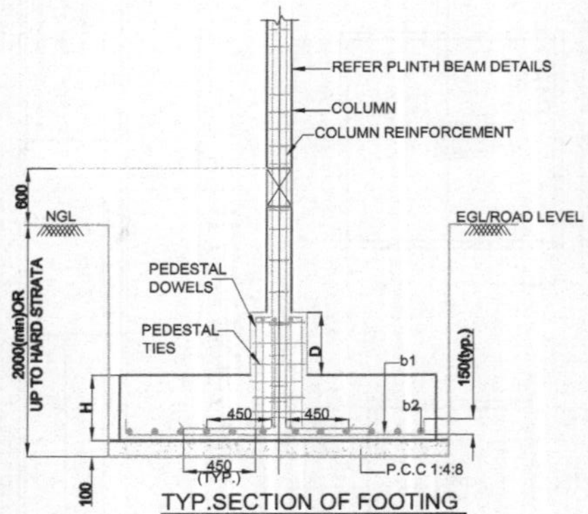

TYP.SECTION OF FOOTING

LEGEND:
EGL - EXISTING GROUND LEVEL
FFL - FINISHED FLOOR LEVEL

COLUMN SCHEDULE (GRADE OF CONCRETE - M25)

COLUMN TYPE	SIZE (B x L)	FOUNDATION TO FIRST SLAB LEVEL	TIES TYPE	TIES
C1	300 x 375	375 x 300; #16-6 Nos.		#8@150 c/c
C2	300 x 375	375 x 300; #20-6 Nos.		#8@150 c/c
C3	300x375	375 x 300; #20-6 Nos.; #16-2 Nos.		#8@150 c/c

FOOTING SCHEDULE

FOOTING	FOOTING SIZE (B x L)	DEPTH	REINFORCEMENT		PEDESTAL SIZE (mm)	PEDESTAL DEPTH (mm)	PEDESTAL REINFORCEMENT	PEDESTAL TIES
		h in mm	b1	b2				
F1	1400x1400	325	#10@150 c/c	#10@150 c/c	525x525	600	#12 -8 Nos	#8@200mm c/c
F2	1950x1950	450	#10@125 c/c	#10@125 c/c	525x525	600	#12 -8 Nos	#8@200mm c/c
F3	2100x2100	475	#10@125 c/c	#10@125 c/c	525x525	600	#12 -8 Nos	#8@200mm c/c
F4	2250x2250	500	#10@125 c/c	#10@125 c/c	525x525	600	#12 -8 Nos	#8@200mm c/c
F5	2550x2550	525	#12@125 c/c	#12@125 c/c	525x525	600	#12 -8 Nos	#8@200mm c/c
F6	2650x2650	550	#12@125 c/c	#12@125 c/c	525x525	600	#12 -8 Nos	#8@200mm c/c

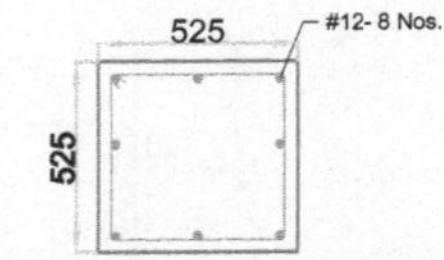

P1 - PEDESTAL 525x525

GENERAL NOTES:

1. GRADE OF CONCRETE MIX SHALL BE AS FOLLOWS
   COLUMNS M-25
   FOOTINGS M-25
   WITH MINIMUM.CEMENT CONTENT OF 320KG/CU.M. UNLESS OTHERWISE STATED.
2. REINFORCEMENT SHALL BE HIGH STRENGTH DEFORMED BARS OF GRADE Fe 500 CONFIRMING TO IS:1786-2008.
3. CLEAR COVER TO REINFORCEMENT AS FOLLOWS:
   FOOTINGS - 2"
   COLUMNS - 1.5"
4. SBC OF SOIL IS 15 T/M^2
5. THIS STRUCTURE IS DESIGNED FOR G+2 FUTURE FLOORS ONLY
6. LAP LENGTH IS 45XD TIMES OF DIA OF THE BAR. (D IS THE SMALLEST DIA OF THE BAR.)
7. LAPS SHALL BE STAGGERED.
10. DO NOT SCALE THIS DRAWING FOLLOW FIGURED DIMENSIONS.
8. LAPPING AT COLUMN BEAM JUNCTIONS SHALL BE AVOIDED.
9. WORK SHALL BE IN ACCORDANCE WITH IS:456 LATEST PRINT.
11. ANY DISCREPANCY SHALL BE BROUGHT TO THE NOTICE OF THE CONSULTANTS.

REV.NO.	DESCRIPTION	DATE	DESIGN	DRAWN
	FOR INFORMATION			

PROJECT:

DRAWING TITILE: COLUMN AND FOOTING SCHEDUE

STRUCTURAL CONSULTANTS:

RELEASED FOR: ☐ CONCEPT ☐ TENDER ☐ SCHEMATIC ☑ APPROVAL ☐ ADVANCE COPY ☐ CONSTRUCTION

DRAWN	REV	SIZE	A3 SIZE
DSGN	VENKAT	REV. NO:	R0
CHKD	PSC	DATE	
APPD	PSC	SHEET NO	4 OF 10
DRG. NO:	- - ST - 004		

Practice Sheet No. 12

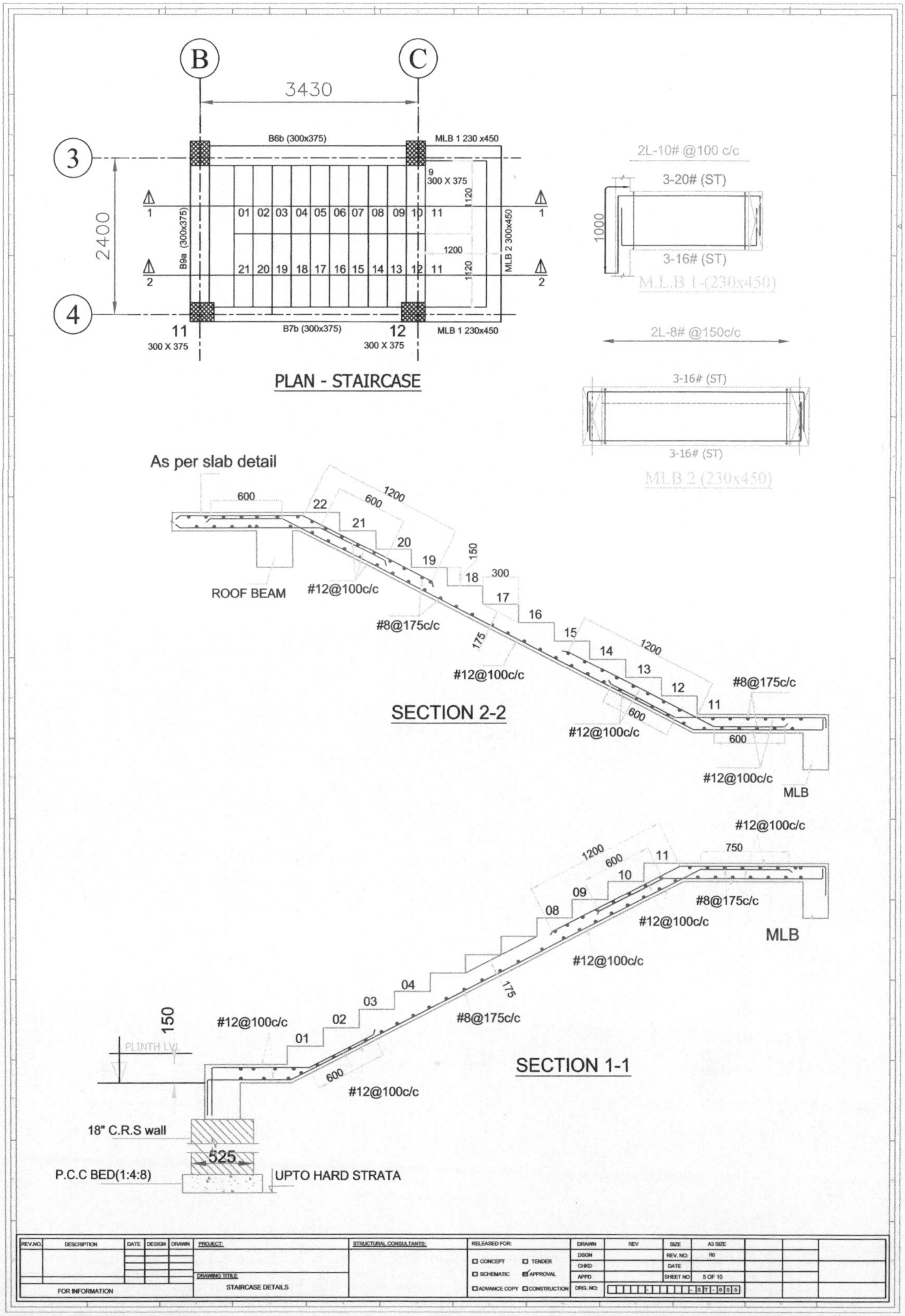

Practice Sheet No. 13

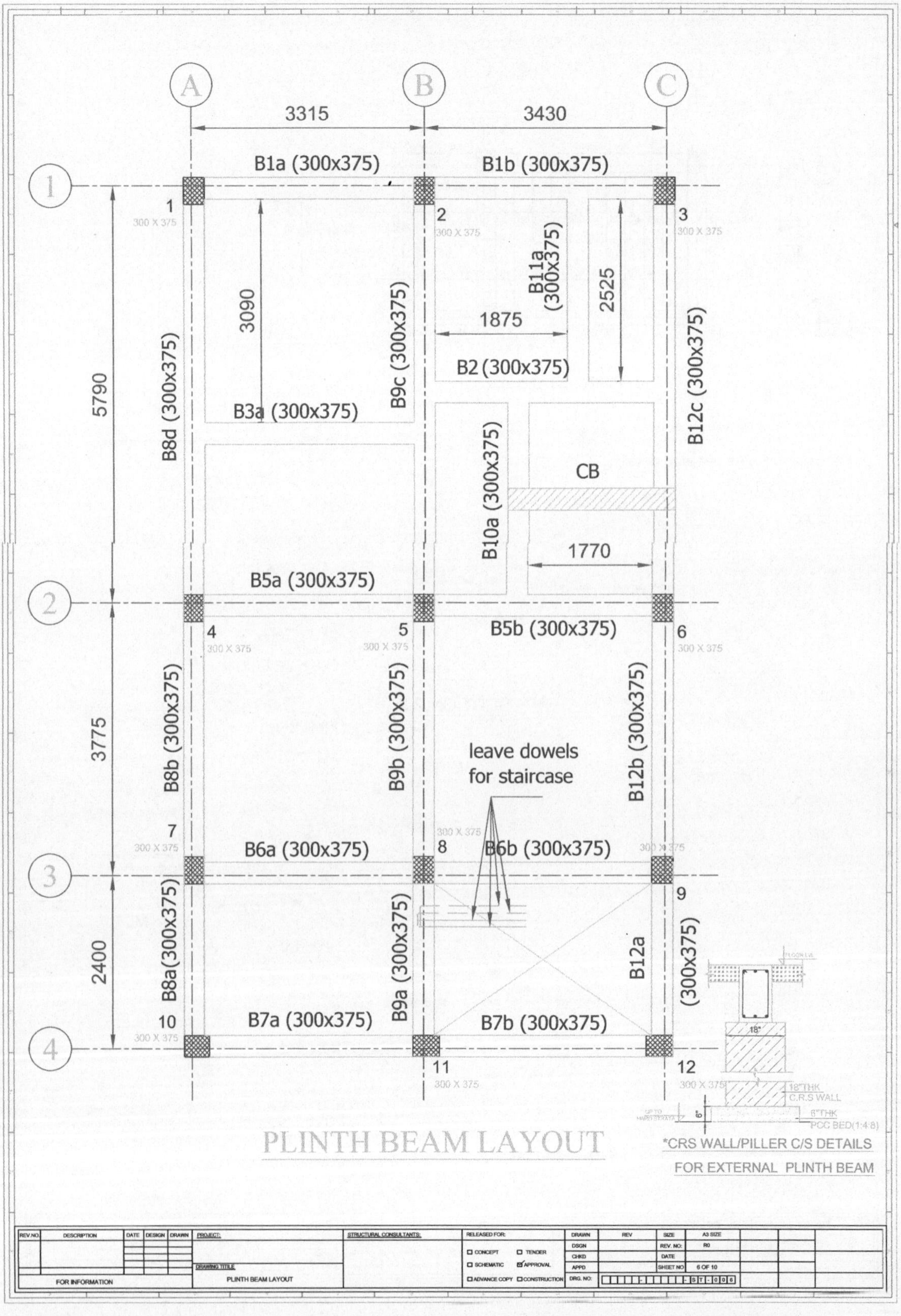

Practice Sheet No. 14

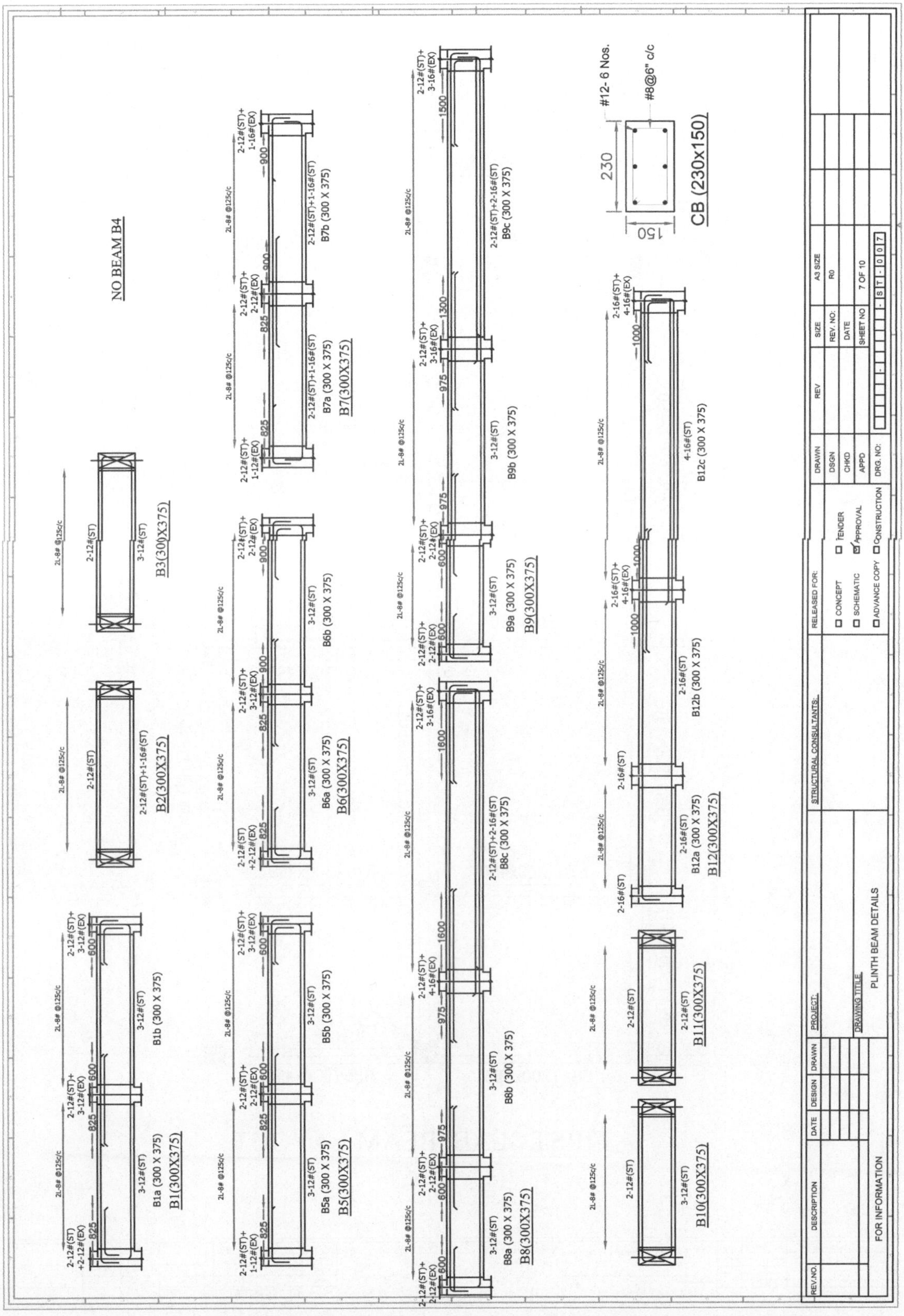

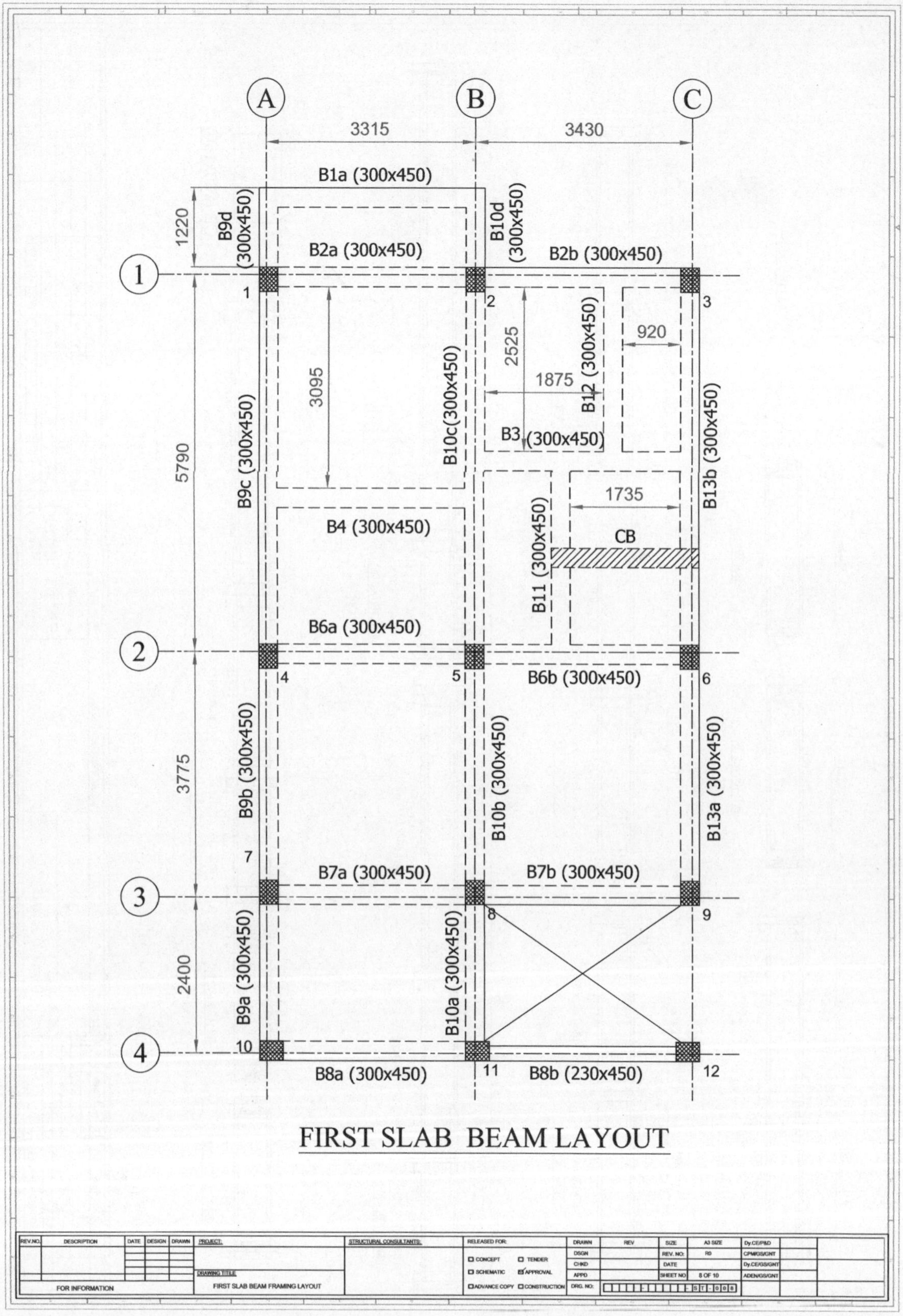
A
B
C
3315
3430
B1a (300x450)
1220
B9d (300x450)
B10d (300x450)
B2a (300x450)
B2b (300x450)
1
2
3
3095
2525
920
1875
B12 (300x450)
B10c(300x450)
B3 (300x450)
5790
B9c (300x450)
1735
B13b (300x450)
B4 (300x450)
CB
B11 (300x450)
B6a (300x450)
4
5
6
B6b (300x450)
3775
B9b (300x450)
B10b (300x450)
B13a (300x450)
7
B7a (300x450)
B7b (300x450)
8
9
2400
B9a (300x450)
B10a (300x450)
10
11
12
B8a (300x450)
B8b (230x450)
FIRST SLAB BEAM LAYOUT
REV.NO.
DESCRIPTION
DATE
DESIGN
DRAWN
FOR INFORMATION
PROJECT:
DRAWING TITLE
FIRST SLAB BEAM FRAMING LAYOUT
STRUCTURAL CONSULTANTS:
RELEASED FOR:
CONCEPT
TENDER
SCHEMATIC
APPROVAL
ADVANCE COPY
CONSTRUCTION
DRAWN
DSGN
CHKD
APPD
DRG. NO:
REV
SIZE
A3 SIZE
REV. NO:
R0
DATE
SHEET NO.
8 OF 10
Dy.CE/P&D
CPM/GS/GNT
Dy.CE/GS/GNT
ADEN/GS/GNT

Practice Sheet No. 16

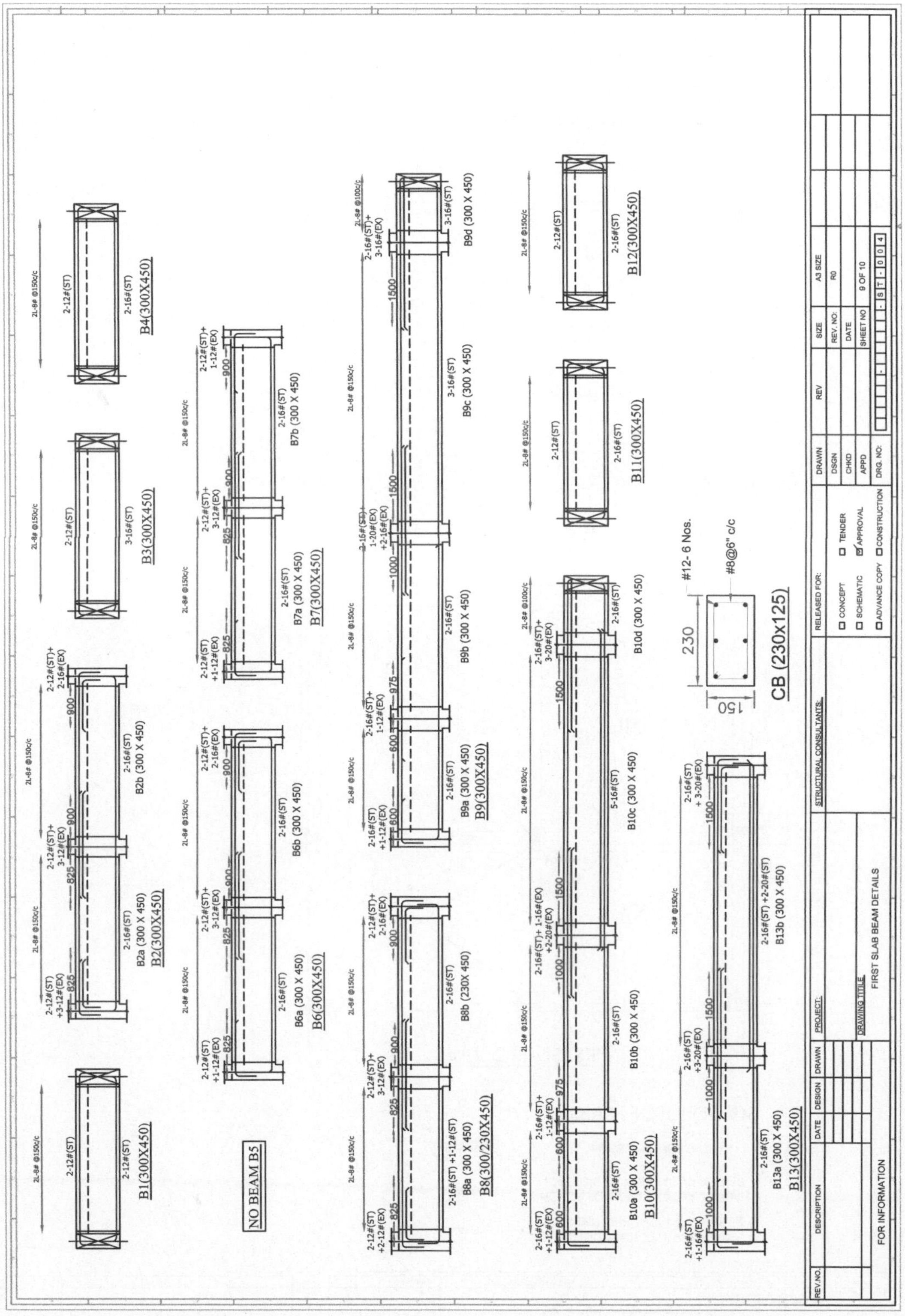

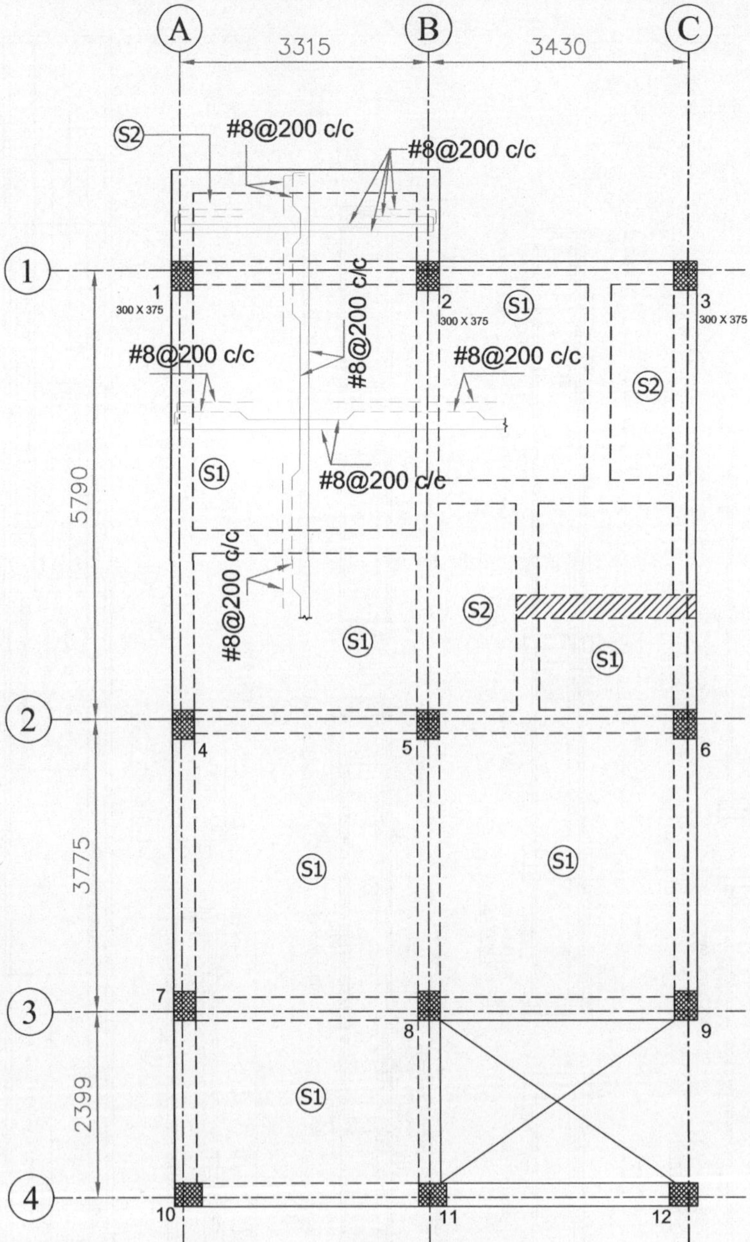

FIRST SLAB DETAILS

SCHEDULE OF SLABS			
NO.S	THICKNESS	SLAB REINFORCEMENT	REMARKS
S1	125	8#@200 c/c IN SHORTER DIRECTION. 8#@200 c/c IN LONGER DIRECTION.	TWO WAY
S2	125	8#@200 c/c IN SHORTER DIRECTION. 8#@200 c/c IN LONGER DIRECTION.	ONE WAY

GENERAL NOTES:

1. GRADE OF CONCRETE MIX SHALL BE AS FOLLOWS SUBJECTED TO MIN. CEMENT CONTENT OF 320 KG/CMT
   SLABS M-25 WITH fck 300 kg/sqcm.
   BEAMS M-25 WITH fck 300 Kg/sqcm.
2. REINFORCEMENT SHALL BE HIGH STRENGTH DEFORMED BARS OF GRADE Fe 500 CONFIRMING TO IS 1786-1985
3. CLEAR COVER TO REINFORCEMENT AS FOLLOWS:
   SLABS -20MM
   BEAMS -25MM
4. CRANK SLAB BARS AT 1/4 SPAN OF CONTINUOUS SUPPORTS 1/7 SPAN FOR DISCONTINUOUS SUPPORTS.
5. EXTEND -VE STEEL UP TO 1/4 SPAN FROM THE FACE ON EITHER SIDE OF THE SUPPORTS FOR BEAMS, AND 1/3 SPAN FOR SLABS FROM THE CENTER OF THE SUPPORTS
6. CURTAIL EXTRA BOTTOM BARS OF BEAMS AT 1/8 SPAN FROM THE FACE OF THE SUPPORTS.
7. READ SLAB THICKNESS AS 125MM, UNLESS OTHERWISE STATED.
8. DISTRIBUTION STEEL 8# @ 200MM C/C UNLESS OTHERWISE MENTIONED.
9. LAP LENGTH 45 X D FOR COMPRESSION
   55 X D FOR TENSION
   (D IS THE SMALLEST DIA OF THE BAR.)
10. LAPS SHALL BE STAGGERED.
11. WORK SHALL BE IN ACCORDANCE WITH IS:456 LATEST PRINT.
12. DO NOT SCALE THIS DRAWING FOLLOW FIGURED DIMENSIONS.
13. ANY DISCREPANCY SHALL BE BROUGHT TO THE NOTICE OF THE CONSULTANTS.
14. READ SLAB REINF. AS 8# UNLESS SPECIFIED.

REV.NO.	DESCRIPTION	DATE	DESIGN	DRAWN	PROJECT:	STRUCTURAL CONSULTANTS:	RELEASED FOR:	DRAWN	REV	SIZE	A3 SIZE
							☐ CONCEPT ☐ TENDER	DSGN		REV. NO:	R0
					DRAWING TITLE:		☐ SCHEMATIC ☑ APPROVAL	CHKD		DATE	
	FOR INFORMATION				FIRST SLAB DETAILS		☐ ADVANCE COPY ☐ CONSTRUCTION	APPD		SHEET NO	10 OF 10
								DRG. NO:	- - ST - 0 0 10		

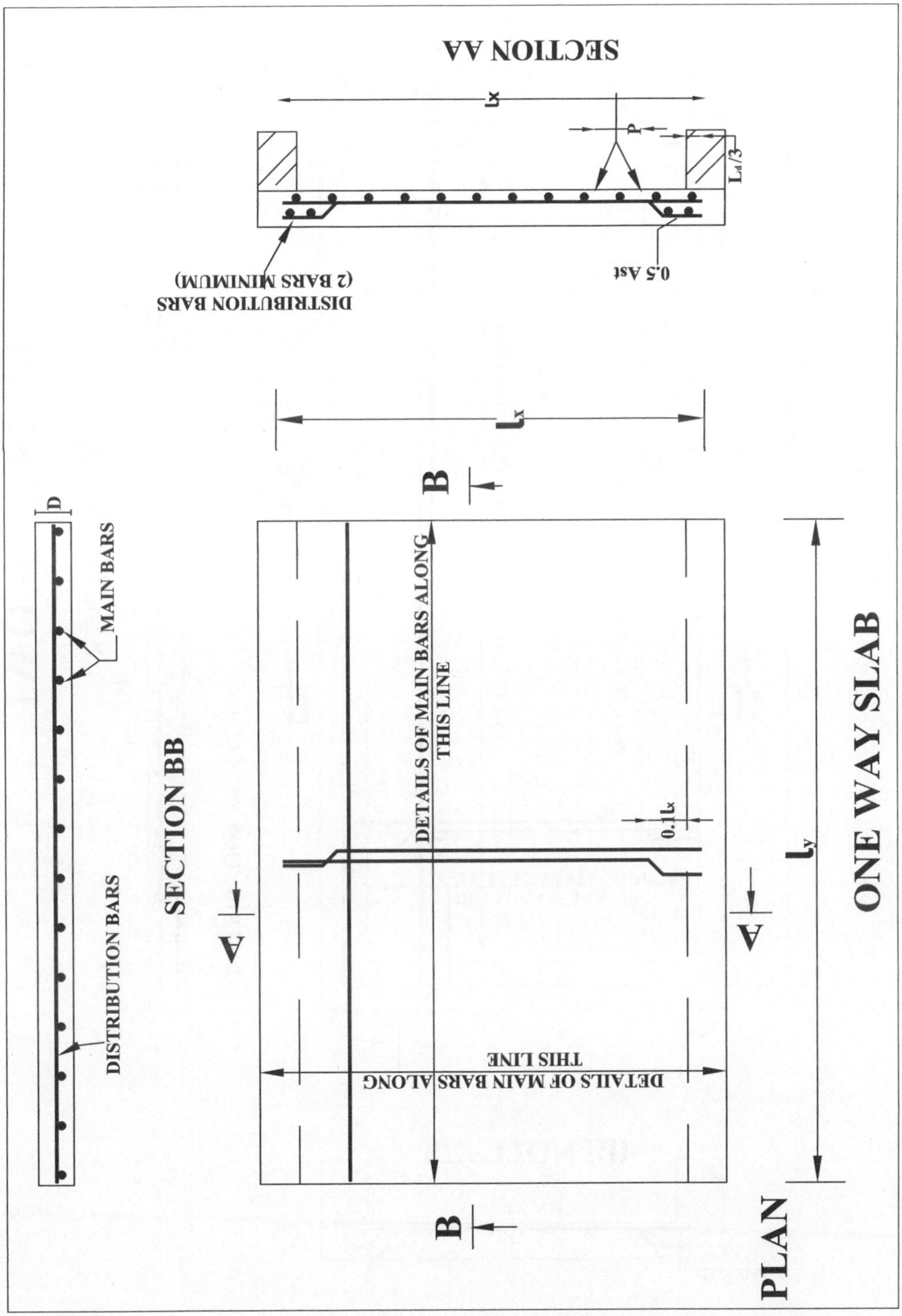
ONE WAY SLAB
PLAN
SECTION AA
SECTION BB
DETAILS OF MAIN BARS ALONG THIS LINE
DETAILS OF MAIN BARS ALONG THIS LINE
MAIN BARS
DISTRIBUTION BARS
DISTRIBUTION BARS (2 BARS MINIMUM)
0.5 Ast
Lx
Ly
0.1Lx
Ld/3
D
A
A
B
B

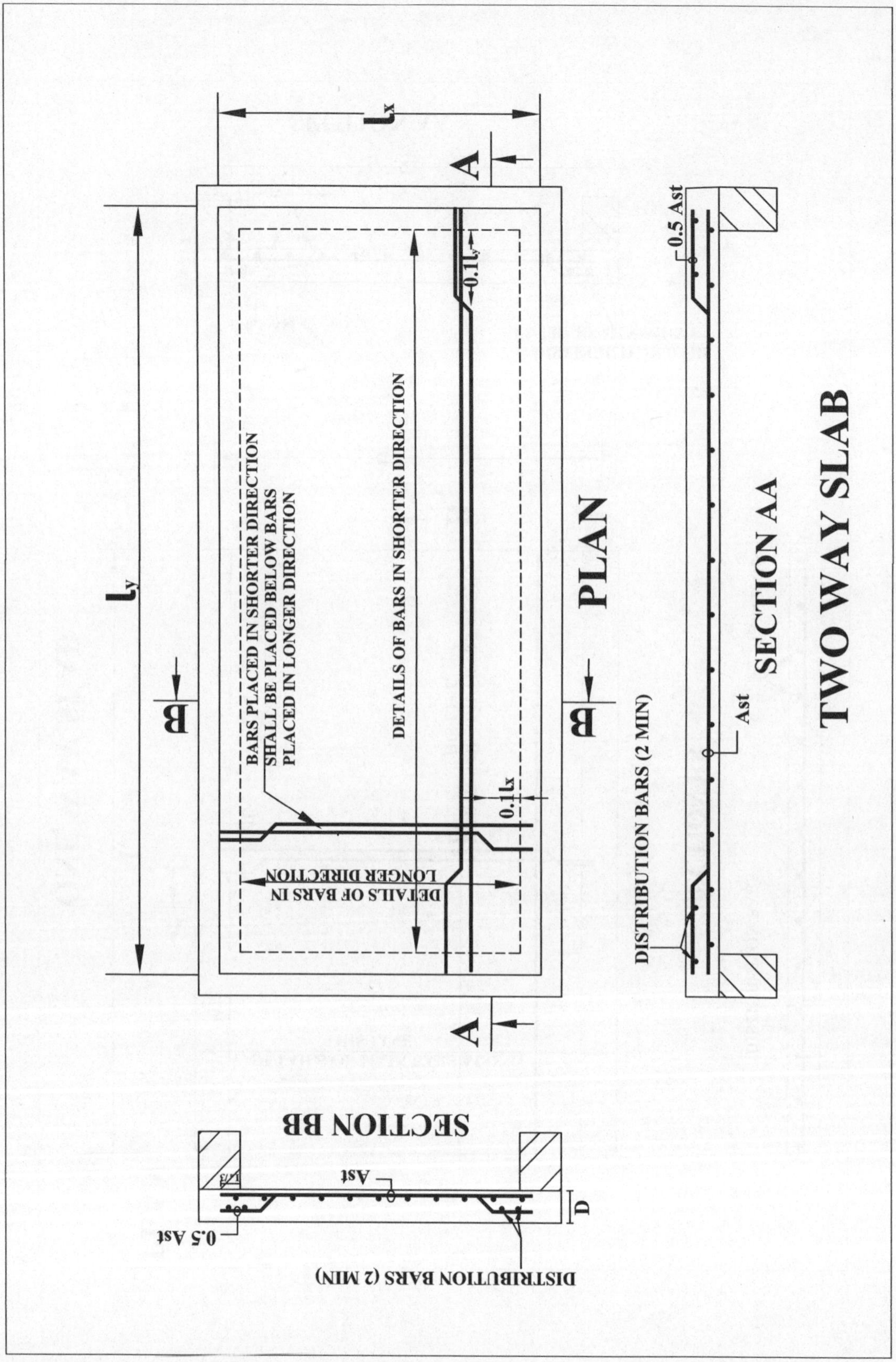
TWO WAY SLAB
PLAN
l_x
l_y
A
A
B
B
BARS PLACED IN SHORTER DIRECTION SHALL BE PLACED BELOW BARS PLACED IN LONGER DIRECTION
DETAILS OF BARS IN SHORTER DIRECTION
DETAILS OF BARS IN LONGER DIRECTION
$0.1l_y$
0.1lx
SECTION AA
0.5 Ast
Ast
DISTRIBUTION BARS (2 MIN)
SECTION BB
0.5 Ast
Ast
D
DISTRIBUTION BARS (2 MIN)

Practice Sheet No. 20

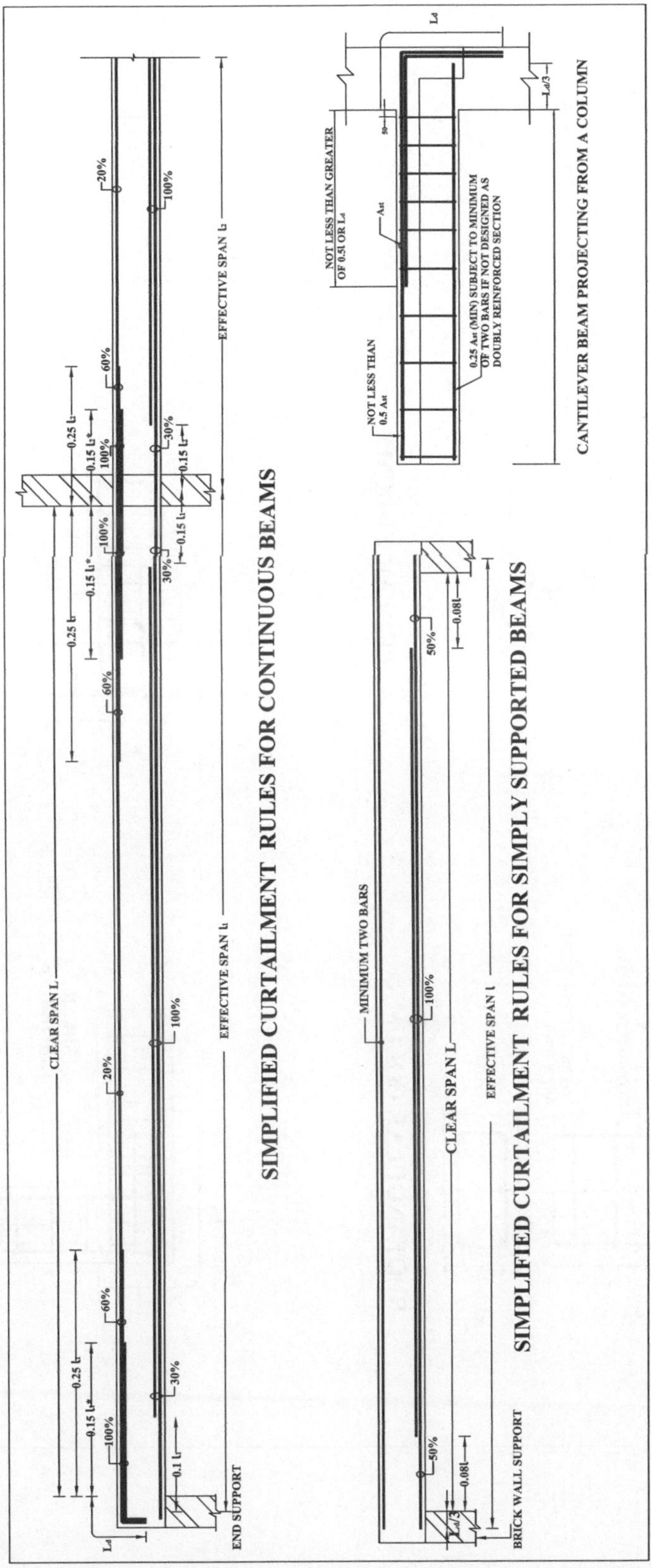

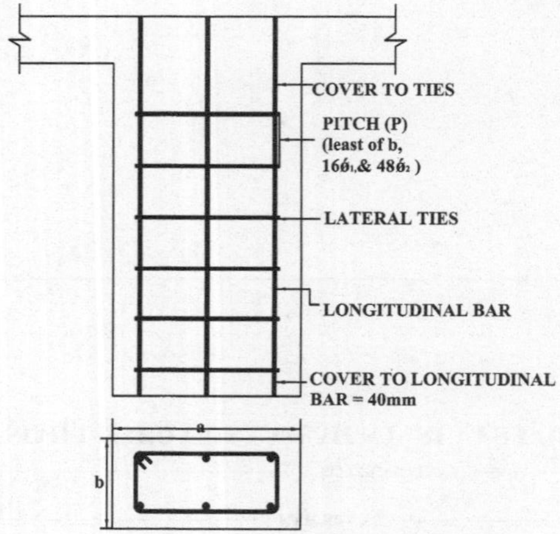

RECTANGULAR COLUMN

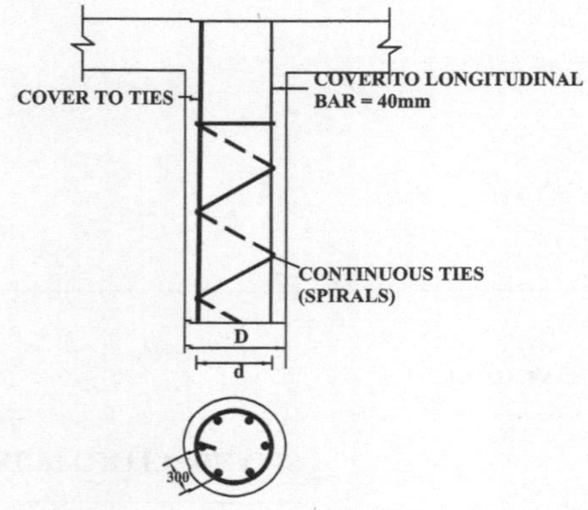

CIRCULAR COLUMN

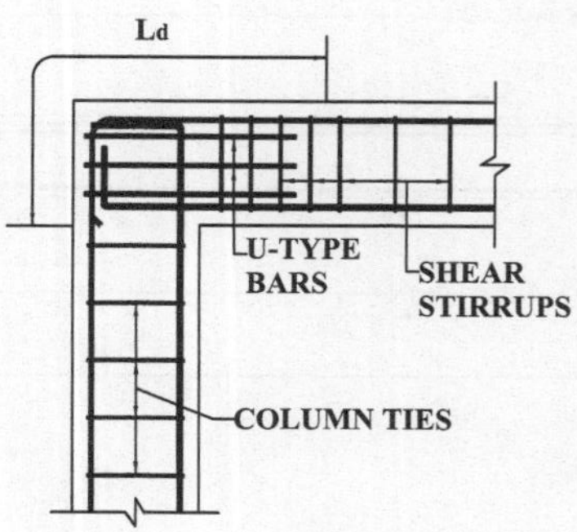

FIXED END POINT IN A COLUMN

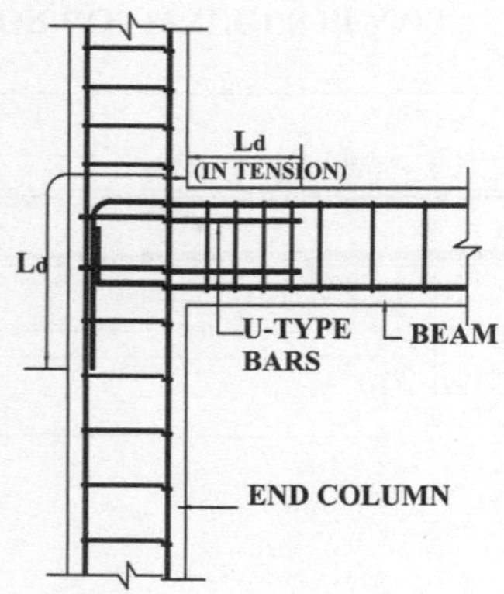

TYPICAL DETAILS OF A BEAM -COLUMN JUNCTION AT EXTERIOR COLUMNS

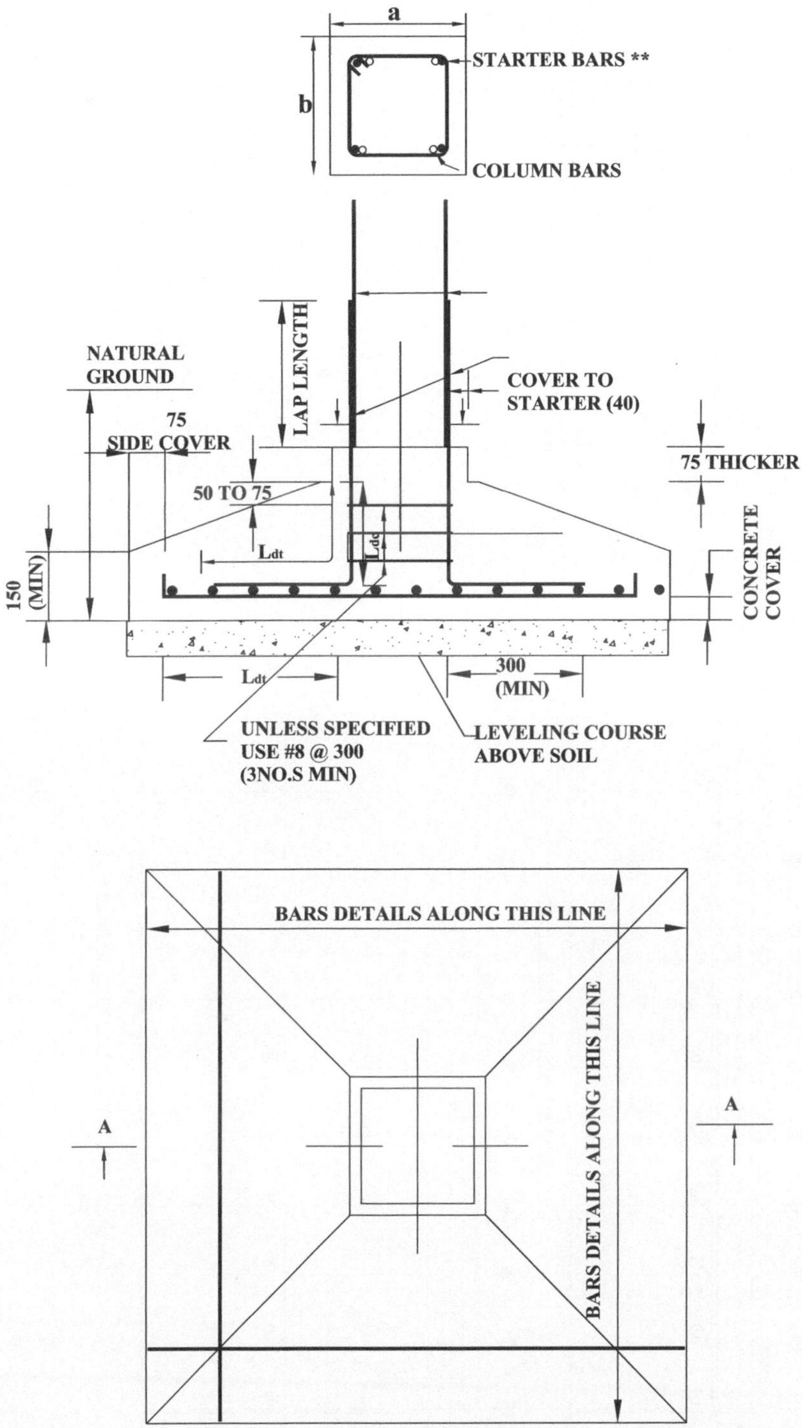

SPREAD FOOTING

Practice Sheet No. 2

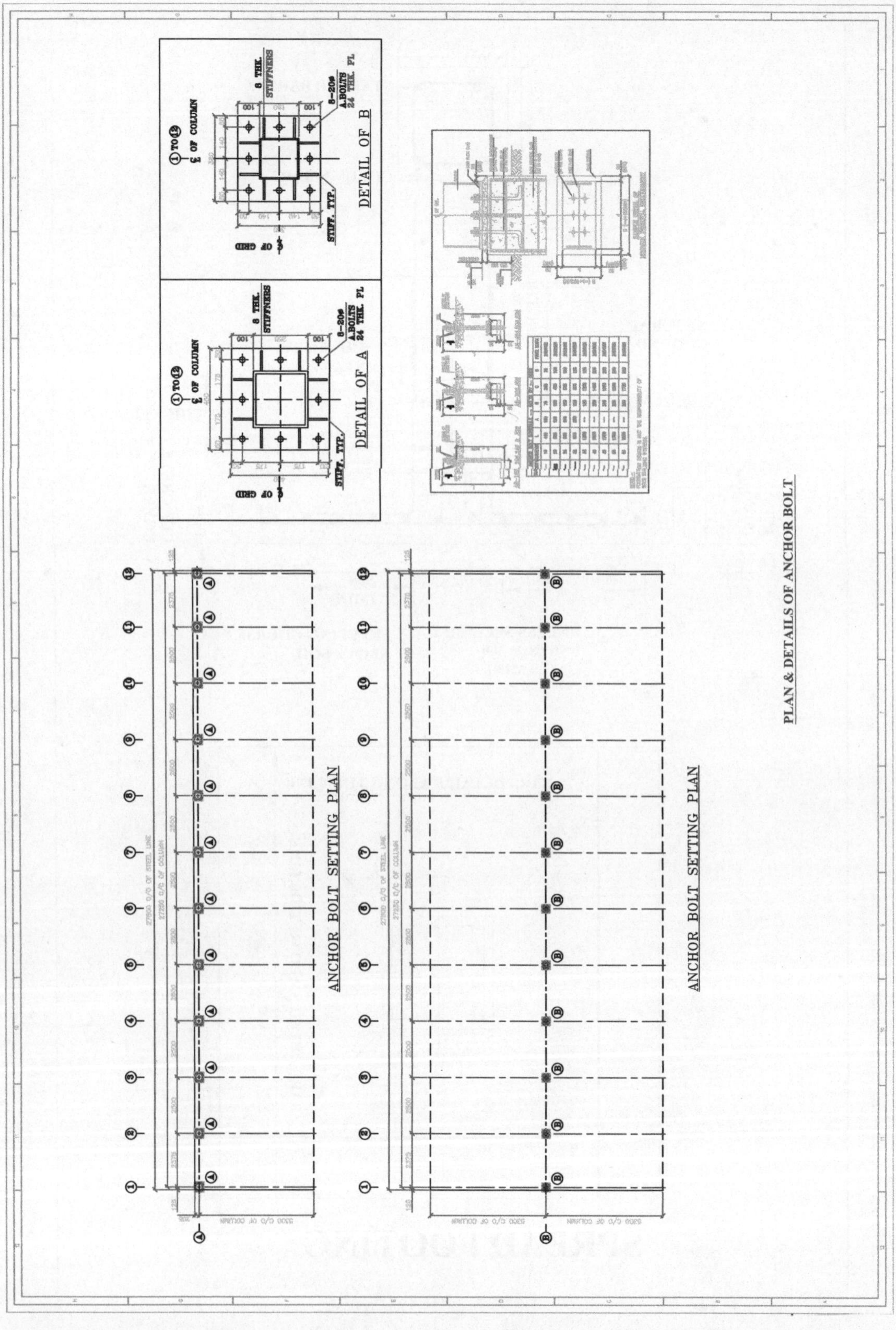

Practice Sheet No. 24

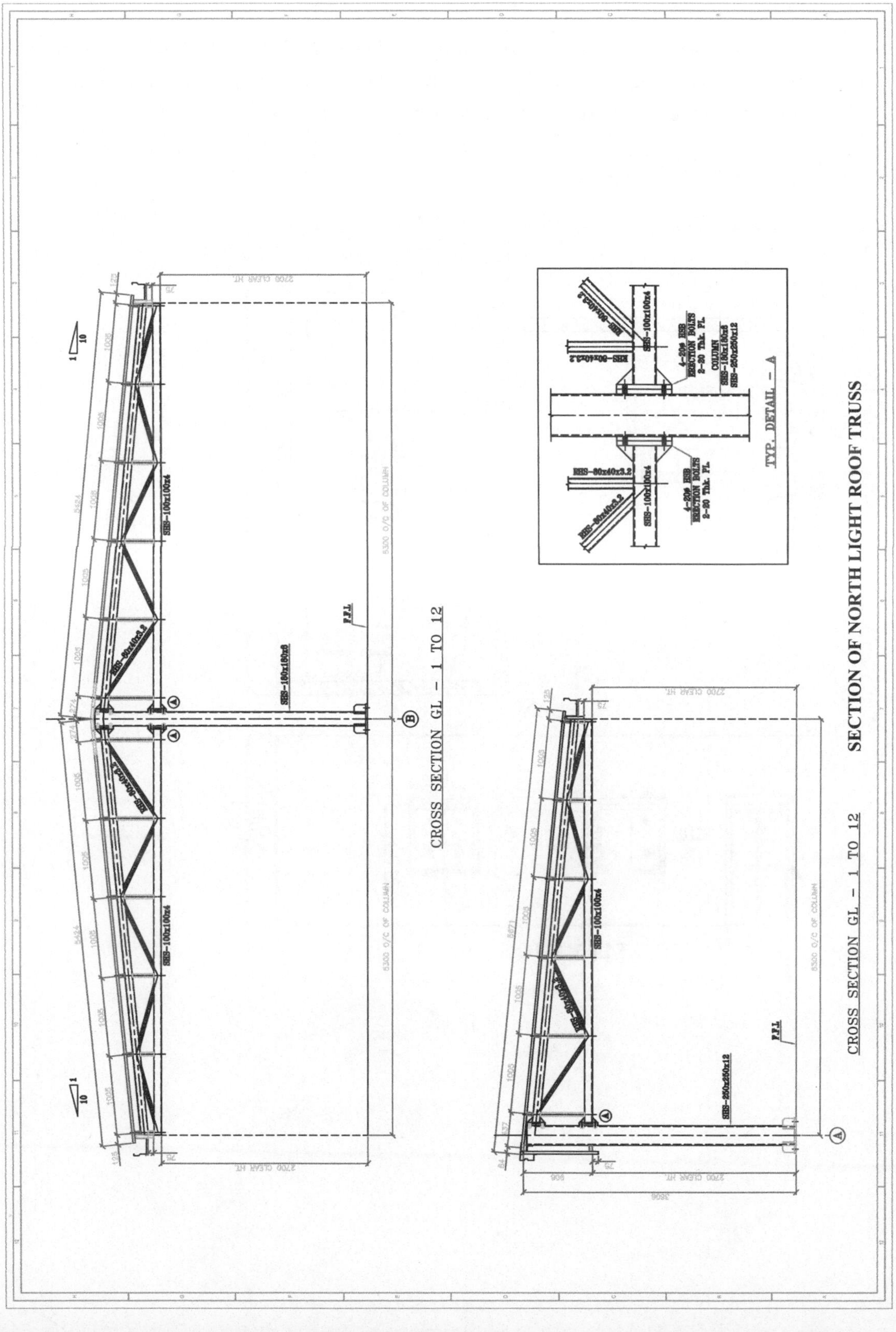

1. Draw the perspective view of the building as shown in the below figure. Assume the Station point and ground level is 10m from the picture plane. Eye point 1.5m above the ground level. Assume 30° as angle of vision.

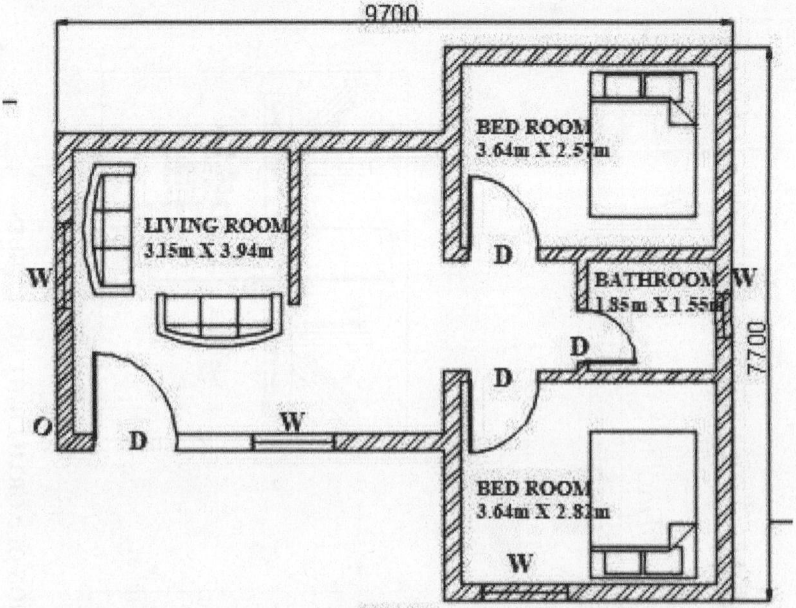

PLAN

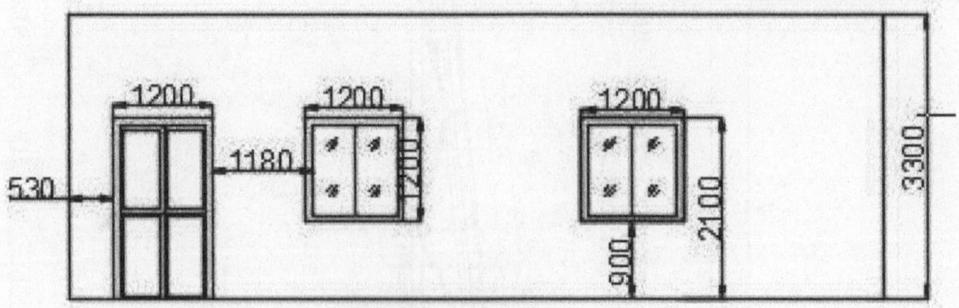

ELEVATION

Practice Sheet No. 25 (b)

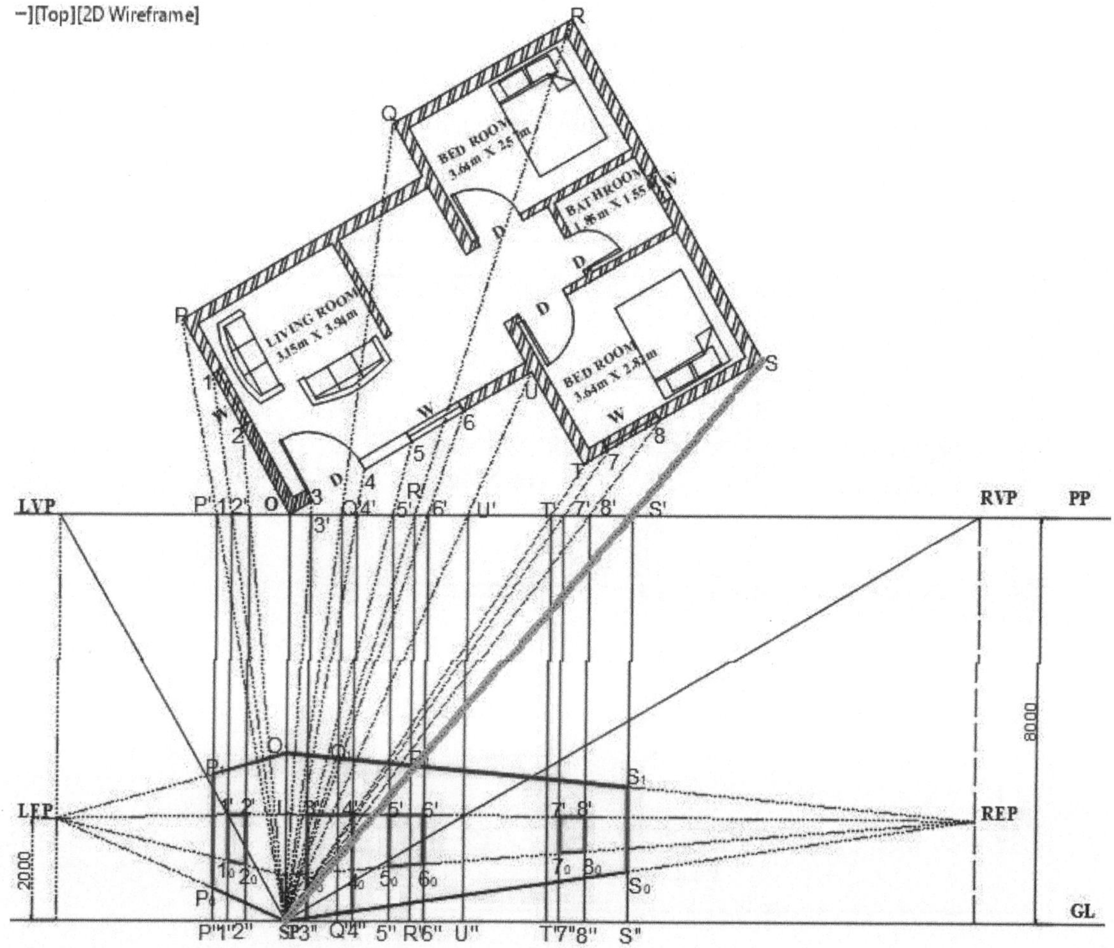

2.Draw the perspective view of the building as shown in the below figure. Assume the Station point and ground level is 8m from the picture plane. Eye point 1m above the ground level. Assume 30° as angle of vision.

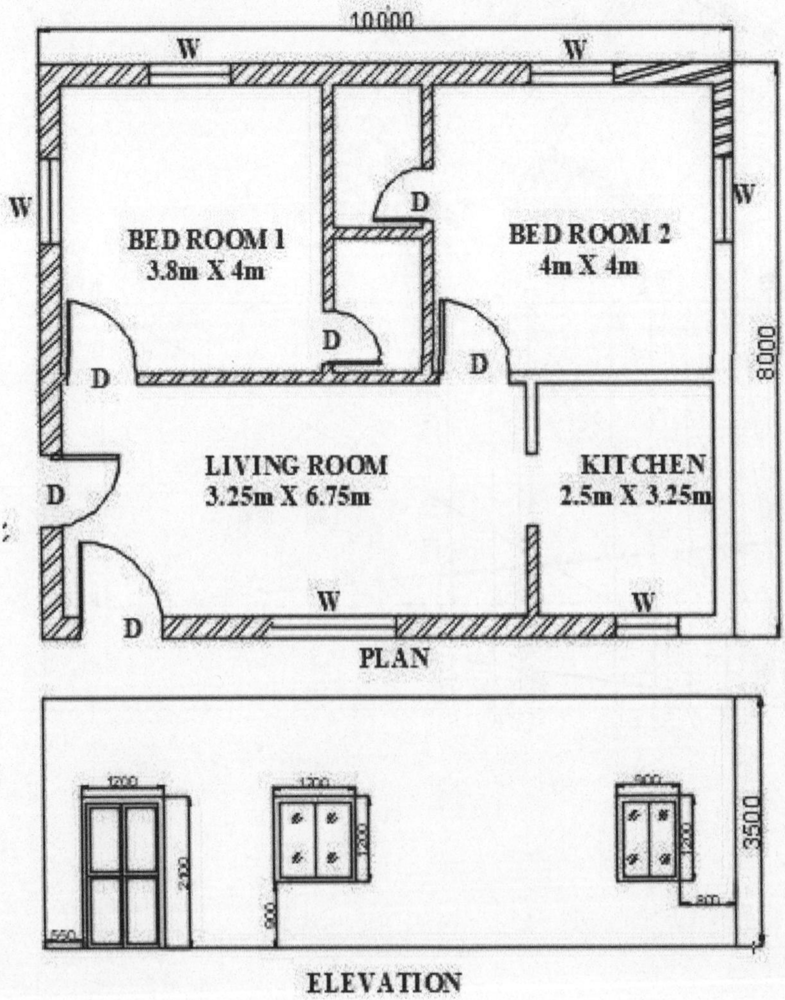

Practice Sheet No. 26 (b)

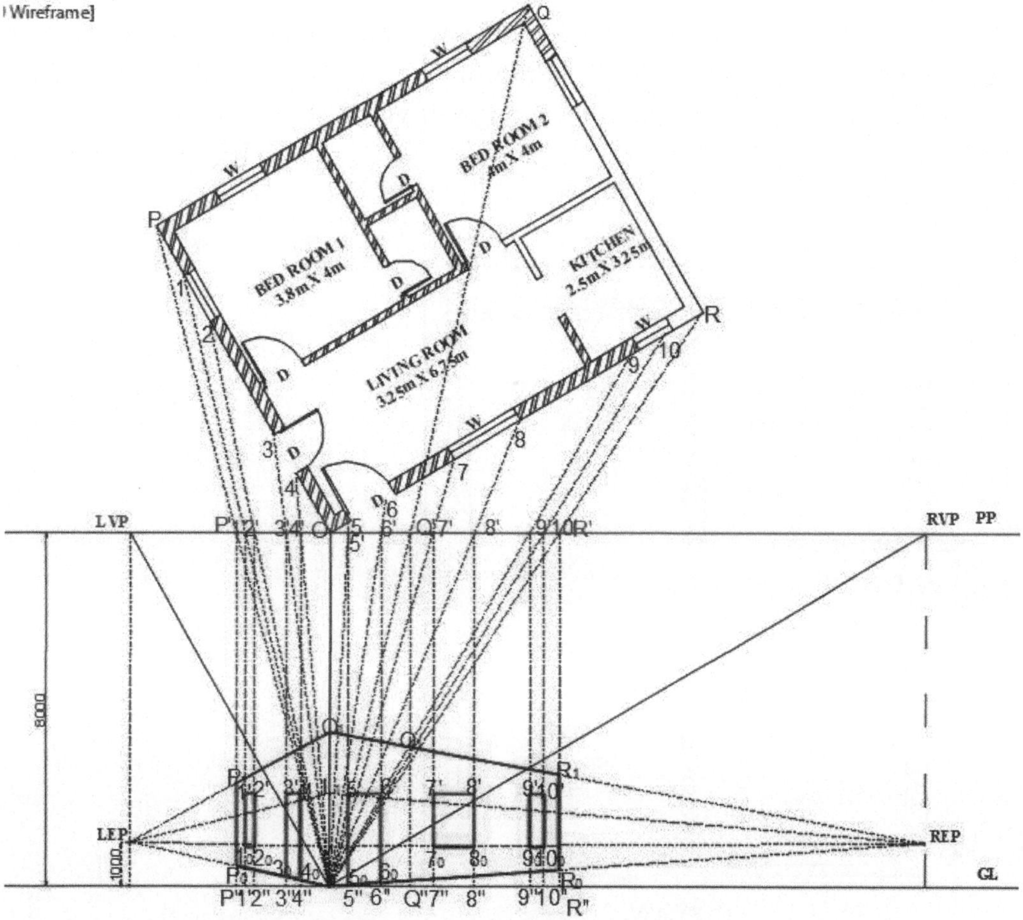

3.Draw the perspective view of the building as shown in the below figure. Assume the Station point and ground level is 9m from the picture plane. Eye point 1.5m above the ground level. Assume 30° as angle of vision.

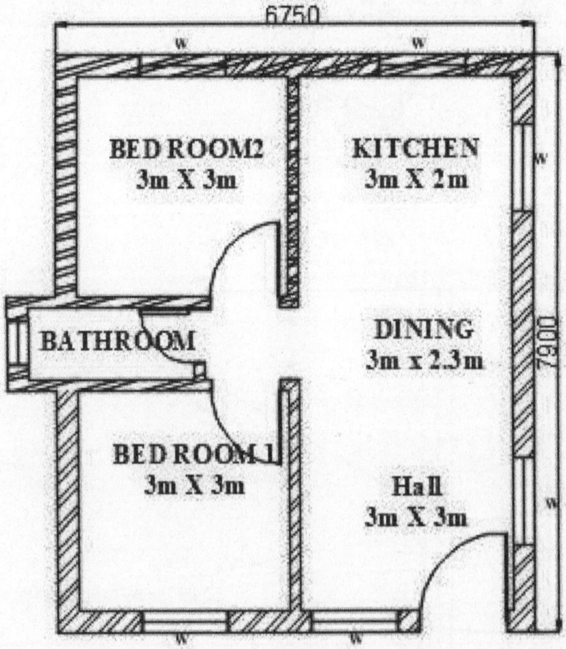

PLAN

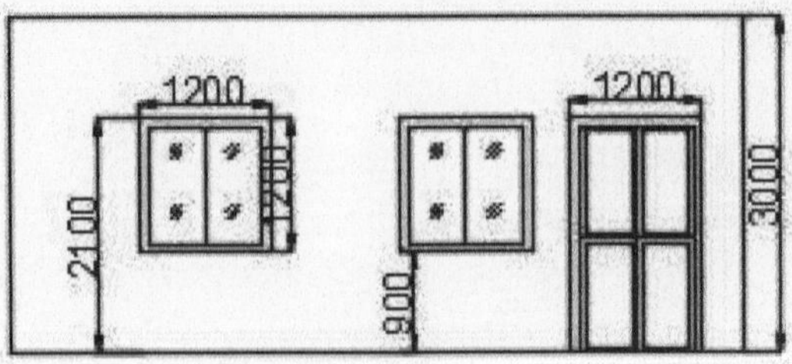

ELEVATION

Practice Sheet No. 27 (b)

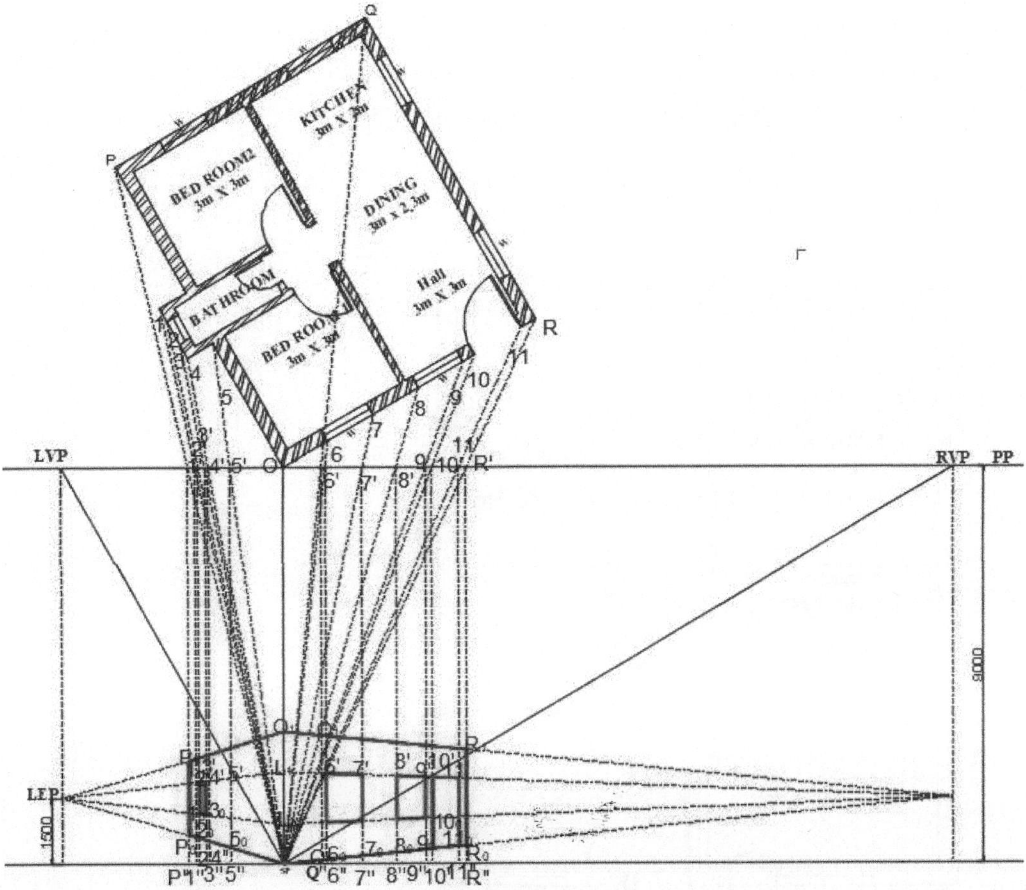

4.Draw the perspective view of the building as shown in the below figure. Assume the Station point and ground level is 10m from the picture plane. Eye point 1.5m above the ground level. Assume 30˚ as angle of vision.

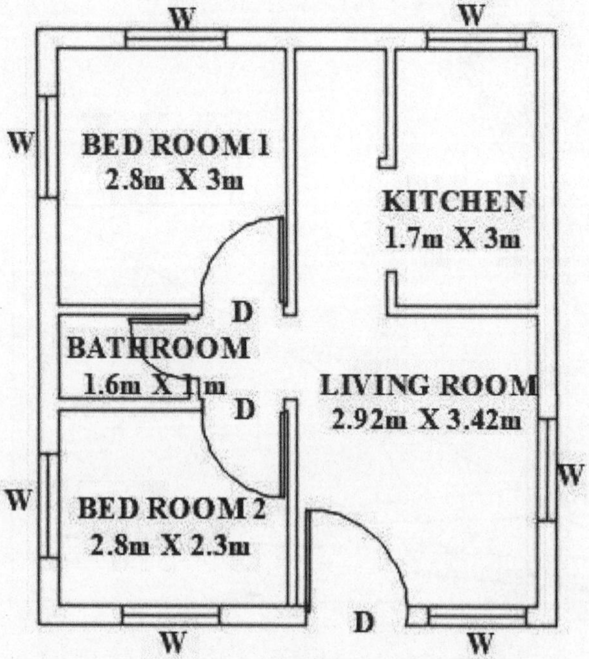

PLAN

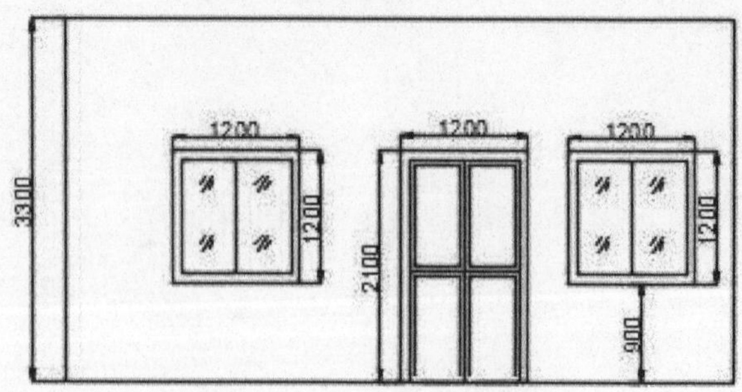

ELEVATION

Practice Sheet No. 28 (b)

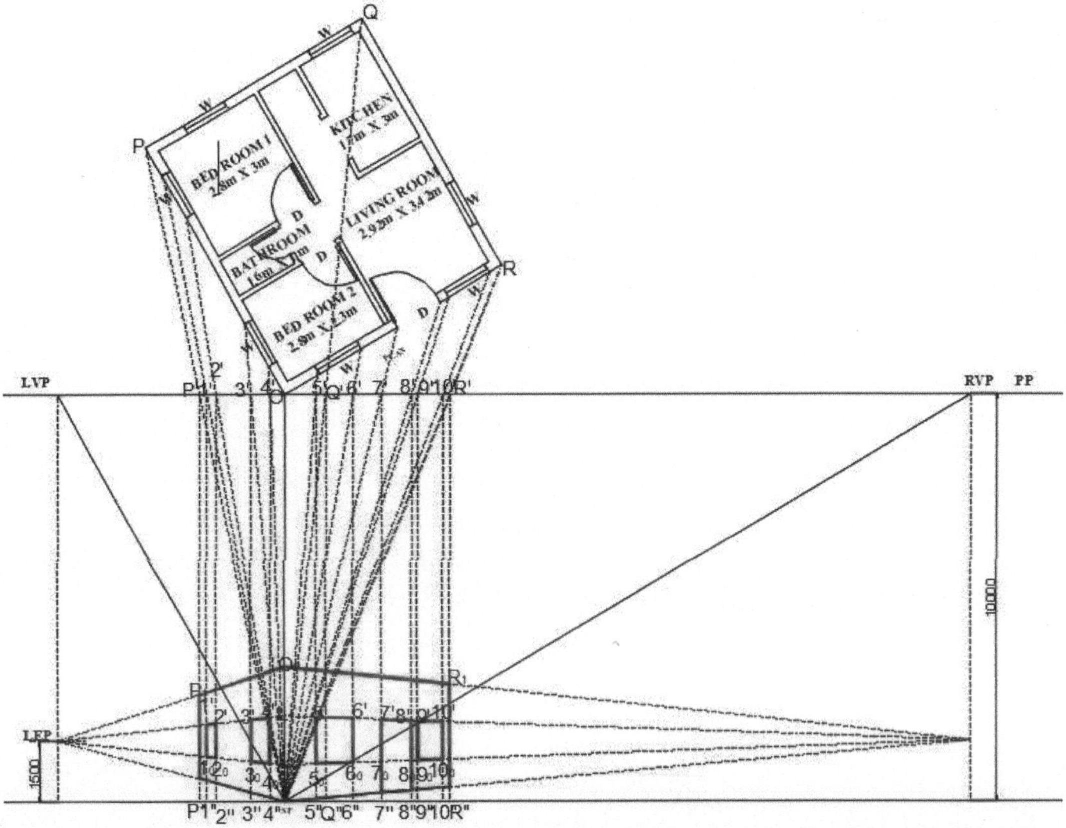